AF581785

Leaching Kinetics of Valuable Metals

Leaching Kinetics of Valuable Metals

Editor

Stefano Ubaldini

MDPI • Basel • Beijing • Wuhan • Barcelona • Belgrade • Manchester • Tokyo • Cluj • Tianjin

Editor
Stefano Ubaldini
Consiglio Nazionale delle Ricerche (CNR)—Istituto di Geologia Ambientale e Geoingegneria (IGAG)
Italy

Editorial Office
MDPI
St. Alban-Anlage 66
4052 Basel, Switzerland

This is a reprint of articles from the Special Issue published online in the open access journal *Metals* (ISSN 2075-4701) (available at: https://www.mdpi.com/journal/metals/special_issues/leaching_kinetics_valuable_metals).

For citation purposes, cite each article independently as indicated on the article page online and as indicated below:

LastName, A.A.; LastName, B.B.; LastName, C.C. Article Title. *Journal Name* **Year**, *Volume Number*, Page Range.

ISBN 978-3-0365-0548-0 (Hbk)
ISBN 978-3-0365-0549-7 (PDF)

Contents

About the Editor

Stefano Ubaldini is a senior research scientist, Member of the International Scientific Advisory Board of the Institute of Geotechnics of the Slovak Academy of Science (SAS), Member of the Board of Directors and of the Scientific Council of the High Tech Recycling International Interuniversity Research Center (HTR), and head of the Bio-Hydrometallurgy, Cyanidation and Electrochemical Laboratories of the Institute of Environmental Geology and Geoengineering of the Italian National Research Council (IGAG-CNR). He is a Member of the Operational Groups, in the framework of the European Innovation Partnership on Raw Materials (EIP-RMs) of Horizon 2020, as a representative of academia, research institutes and think tanks https://ec.europa.eu/growth/tools-databases/eip-raw-materials/en/members. He has been appointed by the European Commission to take part in the Governance of the EIP-RMs as Sherpa member of the High Level Steering Group (HLSG)/Sherpa Group (years 2013–2020).

Relevant skills, scientific activity and experience

Dr. Ubaldini's main area of expertise is in hydrometallurgy and bio-hydrometallurgy. His activity has been carried out mainly in the framework of the primary and secondary raw materials (mineral and materials industry) for over 30 years, including the fundamental research, hydrometallurgy and biometallurgy of low-grade ores, metal extraction from solutions, precious metals recovery with novel leachants, treatment of waste materials, process development and design. His special interest is the interdisciplinary approach (mineralogy, leaching kinetics and solution chemistry) to the problems involved in the extraction and the recovery of metal values from primary and secondary raw materials. He has participated in 15 EC projects and managed R&D projects with funds from industrial and governmental resources; he was the scientist responsible for the following EC projects: H2020, Raw Materials and Recycling, Brite-Euram, VII FP, Life-Environment, INCO-Copernicus and INTAS. Dr. Ubaldini is the author of more than 250 publications on national and international journals and proceedings of international congresses. He has participated, as a component of the scientific committees, international honorific committee, chairman or invited speaker, to numerous national and international conferences. He is Member of International Advisory Board, Guest Editor and referee of international journals in the frame of waste treatment, hydrometallurgy and bio-hydrometallurgy. In the framework of the programme "Cooperation" of the 7th FP and Horizon 2020, he is employed by the European Commission as an expert evaluator. He is an expert evaluator of research projects appointed by the National Center of Science and Technology Evaluation, Ministry of Education and Science, Astana, Republic of Kazakhstan. He is an expert evaluator of the projects of the Italian Ministry of Education, University and Research (MIUR). Dr. Ubaldini is an international evaluator of PhD theses. His current research interests are primary and secondary raw materials, with particular reference versus the hydrometallurgy and biometallurgy of low-grade georesources, with the main aim to recover base and precious metals and to develop sustainable and innovative technologies for the processing of industrial wastes (spent catalysts, industrial tailings, batteries, WEEE etc.) and mining residues, coming also from abandoned sites, and the treatment of waste waters.

Editorial

Leaching Kinetics of Valuable Metals

Stefano Ubaldini

Istituto di Geologia Ambientale e Geoingegneria, CNR, Area della Ricerca di Roma RM 1—Montelibretti—Via Salaria Km 29,300, Monterotondo, 00015 Roma, Italy; stefano.ubaldini@igag.cnr.it; Tel.: +39-06-90672748

Received: 18 January 2021; Accepted: 18 January 2021; Published: 19 January 2021

1. Introduction and Scope

Leaching is a primary extractive operation in hydrometallurgical processing, by which a metal of interest is transferred from naturally-occurring minerals into an aqueous solution. In essence, it involves the selective dissolution of valuable minerals, where the ore, concentrate, or matte is brought into contact with an active chemical solution known as a leach solution.

There are numerous hydrometallurgical process technologies used for recovering metals, such as: agglomeration; leaching; solvent extraction/ion exchange; metal recovery; and remediation of tailings/waste. Currently, hydrometallurgical processes have a wide range of useful applications, not only in the mining sector—in particular, for the recovery of precious metals, such as gold and silver—but also in the environmental sector, for the recovery of toxic metals (such as copper, nickel, zinc, manganese, arsenic, cadmium, chromium, lead) from wastes of various types, and their reuse as valuable metals, after purification.

Therefore, there is an increasing need to develop novel solutions, to implement environmentally sustainable practices in the recovery of these valuable and precious metals, with particular reference to the critical metals, that are those included in materials that are indispensable to modern life and for which an exponential increase in consumption is already a reality or will be in a short-term perspective (antimony, indium, vanadium, rare hearts, etc.). Consequently, the economics of the processes, which are closely linked to the kinetics of leaching, are of great importance.

For publication in this Special Issue, consideration has been given to articles that contribute to the optimization of the kinetic conditions of innovative hydrometallurgical processes—economic and of low environmental impact—applied for the recovery of valuable and critical metals.

I would like to thank the authors who accepted this invitation, helping us to produce a high-impact, high-quality Special Issue on the "Leaching Kinetics of Valuable Metals".

2. Contributions

Researchers around the globe investigating the leaching kinetics of valuable metals have been invited to submit research papers, so that readers can recognize the common points between them. Among the submitted manuscripts, eleven articles have been published in the issue.

The papers are all of high scientific value, while the experimental activities carried out fall into various disciplinary sectors, confirming the importance of the studies of the leaching kinetics of valuable metals in different scientific and technological fields. Here, I will briefly summarize the content of the published papers.

Geochemical characterization studies and batch leaching experiments were conducted to explore the effects of a CO_2 + O_2 leaching system on uranium (U) recovery from ores obtained from an eastern limb of the Zinda Pir Anticline ore deposit in Pakistan [1]. This study provides new insights into the feasibility and validity of the site application of U neutral in situ leaching. According to Asghar F. et al., further studies are needed to reveal the influencing mechanism of the U(VI) initial concentration on U recovery in the solid phase.

While considerable experimental material has accumulated on the purification of uranium-containing water using nanoscale zero-valent iron (nZVI), there are no comparative studies of the sorption properties of iron-containing composites of different composition based on clay minerals. Taking into account the importance of environmental studies based on natural minerals, the removal of uranium by nZVI supported on kaolinite, montmorillonite and palygorskite was investigated by Kornilovych et al., including the removal efficiency of uranium from contaminated groundwater with low and high mineralization [2].

The kinetics of dissolution of refractory sulfide gold-containing concentrates of the Yenisei ridge (Yakutia, Russia) by a solution of HNO_3 in the temperature range of 70–85 °C was investigated by Rogozhnikov et al., leading to the conclusion that the increase in sulfur content in concentrate can be used to ensure the more energy-efficient oxidation of sulfide minerals [3].

A physical–chemical activation desilication process was proposed to extract silica from high alumina fly ash (HAFA) [4]. The effects of fly ash size, hydrochloric acid concentration, acid activation time and reaction temperature on the desilication efficiency, were investigated comprehensively. The results achieved by Gong et al. indicate that physical and chemical activation suppresses the formation of zeolite, thereby improving the desilication efficiency and further improving the A/S of the fly ash; results which are very advantageous for the next step of alumina extraction.

A study of the oxidation of thiosulfate, with oxygen using copper (II) as a catalyst, at a pH between 4 and 5 has been conducted by González-Lara et al. [5]. The basic idea was to avoid the formation of tetrathionate and polythionate, transforming the thiosulfate into sulfate. The nature of the reaction and a kinetic study of thiosulfate transformation, by reaction with oxygen and Cu^{2+} at a ppm level, have been determined and reported. The thiosulfate concentration was reduced from 1 $g{\cdot}L^{-1}$ to less than 20 ppm in less than three hours.

The paper by Batnasan et al. [6] deals with the recovery of gold from waste printed circuit boards (WPCBs) ash by high-pressure oxidative leaching (HPOL) pre-treatment and iodide leaching followed by reduction precipitation. Under the optimal conditions, the percentage of gold extraction from the gold chips and the residue of WPCBs was 99% and 95%, respectively.

The objective of the experimental work carried out by Ubaldini et al. is the application of innovative and sustainable technologies for the treatment and exploitation of mining tailings from Romania [7]. The results obtained by application of the thiosulfate process on a low gold content ore were considered encouraging. The optimization of process parameters and operating conditions should permit the best results in terms of process yields to be achieved.

In the article submitted by Cháidez et al., a copper leaching process from chalcopyrite concentrates using a low-pressure reactor is presented. The experimental results showed that it is possible to extract 98% of copper in only 3 h. This result indicates a fast process compared with others reported in the literature [8].

The experimental results obtained during the preparation of Al-Ni and Al-Ni-Mg alloys using the aluminothermic reduction of NiO by submerged powder injection, assisted with mechanical agitation, are presented and discussed by Silva Beltran et al. [9].

Zhovty Vody city, located in south-central Ukraine, has long been an important center for the Ukrainian uranium and iron industries. Uranium and iron mining and processing activities during the Cold War resulted in poorly managed sources of radionuclides and heavy metals. Widespread groundwater and surface water contamination has occurred, which creates a significant risk to drinking water supplies [10]. The results of the study conducted by Kornilovych et al. demonstrate the effectiveness of the use of the permeable reactive barrier (PRB) for ground water protection near uranium mine tailings storage facility (TSF). The greatest decrease was obtained using zero-valent iron (ZVI)-based reactive media and the combined media of ZVI/phosphate/organic carbon combinations.

Eudialyte is a promising mineral for rare earth elements (REE) extraction due to its good solubility in acid, low radioactivity, and relatively high content of REE. Ma et al. present a study assessing the two stage hydrometallurgical treatment of eudialyte concentrate: dry digestion with hydrochloric acid and leaching with water. The research reported in this paper [11], also as a novelty, explored the feasibility and efficiency of the REE extraction process at room temperature on a scale-up demonstration platform that precedes future industrial applications. Information on upscaling operations for the treatment of eudialyte is also missing in the overall literature.

3. Conclusions and Outlook

A variety of topics have composed this Special Issue, presenting recent developments within the field of leaching kinetics of valuable metals.

As Guest Editor, I am very happy for the success of this Special Issue. I am also proud of the final result, in addition to the high quality and originality of the contributions. I hope that all the scientific results in this Special Issue contribute to the advancement and future development of research in this field.

I would like to warmly thank all the authors for their contributions, and all the reviewers for their efforts in ensuring a high-quality publication. Sincere thanks also to the Editors of *Metals* for their continuous help, and to the *Metals* Editorial Assistants for the valuable and inexhaustible engagement and support during the preparation of this volume. In particular, my sincere thanks to Mr. Toliver Guo for his help and support.

Conflicts of Interest: The author declares no conflict of interest.

References

1. Asghar, F.; Sun, Z.; Chen, G.; Zhou, Y.; Li, G.; Liu, H.; Zhao, K. Geochemical Characteristics and Uranium Neutral Leaching through a $CO_2 + O_2$ System—An Example from Uranium Ore of the ELZPA Ore Deposit in Pakistan. *Metals* **2020**, *10*, 1616. [CrossRef]
2. Kornilovych, B.; Kovalchuk, I.; Tobilko, V.; Ubaldini, S. Uranium Removal from Groundwater and Wastewater Using Clay-Supported Nanoscale Zero-Valent Iron. *Metals* **2020**, *10*, 1421. [CrossRef]
3. Rogozhnikov, D.A.; Shoppert, A.A.; Dizer, O.A.; Karimov, K.A.; Rusalev, R.E. Leaching Kinetics of Sulfides from Refractory Gold Concentrates by Nitric Acid. *Metals* **2019**, *9*, 465. [CrossRef]
4. Gong, Y.; Sun, J.; Sun, S.-Y.; Lu, G.; Zhang, T.-A. Enhanced Desilication of High Alumina Fly Ash by Combining Physical and Chemical Activation. *Metals* **2019**, *9*, 411. [CrossRef]
5. González Lara, J.M.; Cardona, F.P.; Vallmajor, A.R.; Cadevall, M.C. Oxidation of Thiosulfate with Oxygen Using Copper (II) as a Catalyst. *Metals* **2019**, *9*, 387. [CrossRef]
6. Batnasan, A.; Haga, K.; Huang, H.-H.; Shibayama, A. High-Pressure Oxidative Leaching and Iodide Leaching Followed by Selective Precipitation for Recovery of Base and Precious Metals from Waste Printed Circuit Boards Ash. *Metals* **2019**, *9*, 363. [CrossRef]
7. Ubaldini, S.; Guglietta, D.; Vegliò, F.; Giuliano, V. Valorization of Mining Waste by Application of Innovative Thiosulphate Leaching for Gold Recovery. *Metals* **2019**, *9*, 274. [CrossRef]
8. Cháidez, J.; Parga, J.; Valenzuela, J.; Carrillo, R.; Almaguer, I. Leaching Chalcopyrite Concentrate with Oxygen and Sulfuric Acid Using a Low-Pressure Reactor. *Metals* **2019**, *9*, 189. [CrossRef]
9. Beltran, C.S.; Valdes, A.F.; Torres, J.T.; Palacios, R.O. A Kinetic Study on the Preparation of AlNi Alloys by Aluminothermic Reduction of NiO Powders. *Metals* **2018**, *8*, 675. [CrossRef]

10. Kornilovych, B.; Wireman, M.; Ubaldini, S.; Guglietta, D.; Koshik, Y.; Caruso, B.; Kovalchuk, I. Uranium Removal from Groundwater by Permeable Reactive Barrier with Zero-Valent Iron and Organic Carbon Mixtures: Laboratory and Field Studies. *Metals* **2018**, *8*, 408. [CrossRef]
11. Ma, Y.; Stopic, S.; Gronen, L.; Milivojevic, M.; Obradovic, S.; Friedrich, B. Neural Network Modeling for the Extraction of Rare Earth Elements from Eudialyte Concentrate by Dry Digestion and Leaching. *Metals* **2018**, *8*, 267. [CrossRef]

Publisher's Note: MDPI stays neutral with regard to jurisdictional claims in published maps and institutional affiliations.

Article

Geochemical Characteristics and Uranium Neutral Leaching through a CO_2 + O_2 System—An Example from Uranium Ore of the ELZPA Ore Deposit in Pakistan

Fiaz Asghar [1,2], Zhanxue Sun [1,2,*], Gongxin Chen [1,2], Yipeng Zhou [1,2], Guangrong Li [2,3], Haiyan Liu [1,2] and Kai Zhao [2,3]

1 School of Water Resources and Environmental Engineering, East China University of Technology, Nanchang 330013, China; L2017008103@ecut.edu.cn (F.A.); gxchen@ecit.edu.cn (G.C.); 200760037@ecut.edu.cn (Y.Z.); hy_liu@ecut.edu.cn (H.L.)

2 State Key Laboratory of Nuclear Resources and Environment, East China University of Technology, Nanchang 330013, China; liguangrong0086@ecit.edu.cn (G.L.); 2017070914@ecut.edu.cn (K.Z.)

3 School of Geosciences, East China University of Technology, Nanchang 330013, China

* Correspondence: zhxun@ecut.edu.cn; Tel.: +86-79-183-897-597

Received: 11 October 2020; Accepted: 27 November 2020; Published: 1 December 2020

Abstract: Geochemical characterization studies and batch leaching experiments were conducted to explore the effects of a CO_2 + O_2 leaching system on uranium (U) recovery from ores obtained from an eastern limb of Zinda Pir Anticline ore deposit in Pakistan. The mineralogy of the ore was identified by Electron Probe Micro-analyzer (EPMA) and Scanning Electron Microscope-Energy Dispersive Spectrometer (SEM-EDS), showing that pitchblende is the main ore mineral. XRD was also used along with EPMA and SEM characterization data. Experimental results indicate that U mobility was readily facilitated in the CO_2 + O_2 system with Eh 284 mV and pH 6.24, and an 86% recovery rate of U_3O_8 was obtained. U speciation analysis implied the formation of $UO_2(CO_3)_2^{2-}$ in the pregnant solution. The plausible mechanism may be attributed to the dissolved CO_2 gas that forms carbonate/bicarbonate ion releasing oxidized U from the ore mineral. However, U recovery in the liquid phase was shown to decrease by higher U(VI) initial concentration, which may be due to the saturation of Fe adsorption capacity, as suggested by an increase in Fe concentration with increasing initial U(VI) concentration in the solid phase. However, further studies are needed to reveal the influencing mechanism of U(VI) initial concentration on U recovery in the solid phase. This study provides new insights on the feasibility and validity of the site application of U neutral in situ leaching.

Keywords: geochemical characteristics; pitchblende; U neutral leaching; ELZPA ore deposit in Pakistan

1. Introduction

Uranium (U) ore deposits are important to society as a primary material for the generation of nuclear power [1–3]. Interest in in situ leach (ISL) mining has grown considerably over the last 30 years [4] because of its significant advantages over conventional methods for mining, especially low-grade sandstone-type ore deposits. Research on U mobilization has focused on its migration and mineralization under natural conditions, as well as its impact on the environment [5–17]. The incorporation or release of U to or from mineral structures is determined primarily by its redox state and mineral solubility [18]. The change in oxidation state from U(IV) to U(VI) usually occurs in nature under oxidizing conditions at the geosphere/atmosphere interface or within the lithosphere from contact with oxygenated groundwater [19]. Generally, sulfuric acid/iron sulfate and carbonate/bicarbonate are common leachants in acidic and alkaline leaching systems, respectively [20,21].

A number of leaching tests have been reported to study the U leaching mechanism from porous media under different leaching conditions [22–27]. The impact of parameters including pH, Eh, and alkalinity on U recovery parameter was investigated. Combing a modeling approach, U chemical leaching and adsorption tests were performed by Briganti et.al. [28], showing that pH and alkalinity were main factors regulating geochemical behavior of U during ignimbrite-water interaction. However, their study lacks the consideration of Eh and suggests that the greatest leaching of U occurs at 50 °C, which seems irrational in field trials [29,30]. Furthermore, U leaching efficiency was shown to be complicated by factors including chemicals' initial concentrations and Fe and other metals' adsorption, of which the influential mechanisms are not sufficiently studied.

The influential ISL parameters of U recovery can be classified into two different categories. One concerns technological factors and the other concerns natural factors where U ores are leached mostly under oxidizing conditions in order to convert U contained into minerals from a relatively insoluble tetravalent form (U(IV)) into water-soluble hexavalent U (U(VI)) [31]. Under the $CO_2 + O_2$ leaching condition, their partial pressure was kept constant for U dissolution. The constant pressure parameter may produce in situ conditions during laboratory experiments. Indeed, the dissolution of U minerals is promoted by increasing the partial pressure of O_2 or CO_2, which is covered in a later discussion. Hence, it is essential to describe redox conditions and solution chemistry in the natural prevalent environments in order to better understand the geochemical behavior of the U species.

The presence of the U species in a water-rock interaction system is attributed to the evolution of redox conditions. Although acidic and alkaline leaching of U from limestone and sandstone media has been practiced extensively in the past, the shortcomings presented the opportunity to investigate the use of neutral leaching of U to understand its leaching mechanism, considering that limited research has been published on U neutral leaching. The novelty of this work is to unravel the mechanism of U mobilization during the neutral leaching process from U ore of an eastern limb of Zinda Pir Anticline (ELZPA) ore deposit. Therefore, the objectives of this study were the following: (i) identify the deposit minerology and geochemical behavior of the U species and (ii) determine the effects of leaching parameters on the release of U from U-bearing sandstone.

2. Materials and Methods

2.1. Geological and Hydrological Setting of the Study Deposit

The study deposit is located within the eastern ward extension of the Sulaiman Range, Dera Ghazi Khan, Pakistan, known as the eastern limb of Zinda Pir Anticline (ELZPA) ore deposit. It is a typical sandstone-type U deposit. The exposed rocks in the Sulaiman Fold Belt range were deposited from the Triassic to Tertiary periods with an estimated thickness of more than 7000 m [32–36].

The ore body is mainly found at a burial depth of 70–120 m in the Neogene molasses sequence, commonly known as Siwalik, that was uplifted and deformed to form the Zinda Pir Anticline range. The strata dip steeply to the east with angles of between 75° and 80°. The layers above and below the mineral-bearing strata are impermeable and retain confined pore water at varying water table depths. The water has a pH between 7 and 8, and consists mainly of SO_4^{2-}, HCO_3^-, Na^+, Ca^{2+}, and Mg^{2+} ions, as shown in Table 1.

The ELZPA deposit is divided into four zones (locally), i.e., Lal-Ashab uraniferrous horizon (Z-I), Fowl creek uraniferrous horizon (Z-II), Zamdani uraniferrous horizon (Z-III), and Ghazzi uraniferrous horizon (Z-IV). All the zones except Z-I are bounded by lower and upper shales, and a volcanic ash layer deposited parallel to zone IV. The ore body (Z-I) is tabular upward and exhibits a slanting shape along the down dip. The host sandstone is overlain by 7 to 10 m thick brownish grey shale, running parallel to the strike of the host sandstone. The shale below the ore-bearing strata is however not continuously developed. The three predominant formations that form the major subdivision of the Siwalik series are the Vihowa, Litra, and Chaudhwan Formations. The deposit is located in the middle part of the Litra Formation (Middle Siwalik). The hosted sandstone is grey to dirty grey, fine- to medium-grained and loosely to moderately cemented. Sometimes it is cross-bedded and contains limonitic and hematitic alternation bands. The sandstones in the lower part are moderately cemented, thus forming high peaks, whereas those in the upper part are less cemented and fragile, thereby forming a flatter topographic scenario. The average permeability coefficient and porosity of the rock aquifer are 0.61 m/d and 22%, respectively. The climate in the area is semi-arid to arid. The maximum and the minimum annual temperatures are 50 and 5 °C (average temperature is 30 °C). A frost period rarely occurs except for some seasonal snowfall on high peaks of the Sulaiman Range. The precipitation is about 200 mm, which falls mostly during the period of July to September and winter [37,38].

Table 1. Chemical composition of ground water (ppm).

Sample	pH	Ca^{2+}	Mg^{2+}	Na^{+}	K^{+}	HCO_3^-	SO_4^{2-}
Z2TW1	7.35	144	57	200	6	195	619
Z2TW2	7.11	145	101	175	5	176	1122
Z2E-148	7.46	130	102	300	5	171	978
T1WNE2	7.46	64	46	163	4	152	293
FR1T-826C	7.88	51	33	75	4	190	321
Z2TW28/88	7.80	139	21	202	9	156	118
Z2TW26/77	7.68	152	71	220	10	166	1163
Z2TW23/68	7.80	166	123	312	9	156	1274
E-312	7.90	81	68	168	7	207	651
Z1W2230	7.80	90	43	176	7	171	665
Z1W2247	6.75	80	40	245	12	249	574
E-324	8.17	211	113	180	9	130	1302
Z2TW27/42	7.70	142	91	225	12	192	1206

2.2. Sample Collection

The material was sourced from the sandstone-type U deposit that existed in the ELZPA, Sulaiman Range, Dera Ghazi Khan, Pakistan. At each sampling location, three replicate samples were collected and the unrepresentative materials were removed. The depth of the investigated boreholes spanned from 90 to 165 m. Samples were coated with wax during the core recovery process to avoid contact with atmospheric oxygen. All samples were transported to the laboratory for further treatment (discussed below).

2.3. Sample Preparation

Representative (in composite) ore samples were prepared before being analyzed and for laboratory leaching experiments. Briefly, the natural grain size of the crushed ore sample was used for the leaching batch test. For this study, natural grain size means that the ore sample was first gently loosened without reducing the sand grain size. The corresponding particle size distribution ranges were <0.5 mm, 0.5 mm to 1.00 mm, and >1.00 mm. A total of 36 samples were dried, homogenized, and sieved. All the ore samples were grouped as mixed, high, medium, and low grade on the basis of % U_3O_8 concentration, as shown in Table 2. The coning and quartering method was used to prepare representative samples. A portion of the homogenized ore material was then milled to 200 mesh size for the determination of % U_3O_8 in each sample group.

2.4. Reagents

All chemicals used in this work were of analytical reagent grade and used without further purification. The solutions and dilutions used for calibration were prepared using deionized water. Lixiviant solutions used for batch leaching experiments were prepared with laboratory tap water

(LTW), of which the composition is shown in Table 3. The representative ore sample was first grounded into a final 200 mesh size and pulverized before being analyzed (discussed below).

Table 2. Sample grouping.

Sample	Group Name	% U_3O_8 Range
TCS-04	Mixed	0.01 to all
TCS-03	High	0.051 and above
TCS-02	Medium	0.036 to 0.050
TCS-01	Low	0.01 to 0.035

Table 3. Chemical composition of laboratory tap water (g/L).

U_3O_8	HCO_3^-	CO_3^{2-}	Ca^{2+}	Mg^{2+}	SO_4^{2-}	Cl^-	Na^+	Fe^{2+}	$Fe_{(T)}$
BDL	0.18	0.02	BDL	BDL	0.03	0.01	0.11	BDL	BDL

BDL—Beyond Detection Limit.

2.5. Batch Leaching Process

Batch extraction tests typically involve mixing a sample with a specific amount of leaching solution without renewal of the leaching solution [39]. Leaching experiments were carried out in batches with a liquid-to-solid (L/S) ratio of 2 in the autoclave reactor [40]. A Karl Kolb Scientific Technical Supplies autoclave (25 bar, BTR temperature of 300 °C, volume of 0.5 L, Dreieich, Germany) was used. An amount of 150 g of ore sample (TCS-04) was mixed in the prepared 300 mL of lixiviant (liquid) of varying conditions (variable bicarbonate solution concentration), while the ore samples (TCS-03 and TCS-02) were processed under only the best result of HCO_3^- concentration. Experiments were performed to investigate the long-term release of the recalcitrant uranium from the ore under varying lixiviant solution conditions (variable bicarbonate solution concentration), and replicate tests were conducted under the same conditions. The leached solution was sampled at regular time intervals, i.e., from 0.5, 2.5, 7.5, 17.5, 41.5 h to days, and so on, until the equilibrium condition was acquired (i.e., no further change in U_3O_8 recovery in the leached solution). Leaching tests performed by the researchers indicated that the L/S ratio and grain size were important factors for determining the metal release; however, 24 h was not sufficient for a thorough assessment of leaching [41]. The leaching process was kept free from mechanical agitation at room temperature. The batch sampling process consisted of removing 10 mL of leached solution and filtering it through a 0.45 µm filter (Millipore, Merck, Germany). The pH and Eh were monitored throughout the batch process. The filtered sample was analyzed for U_3O_8 and HCO_3^- concentration. Immediately after sampling, the sample was kept in a refrigerator until analysis. The HCO_3^- concentration was controlled by adding ammonium bicarbonate (NH_4HCO_3) in the lixiviant solution. The pH of the leaching regulation system was set in the range of 6 to 6.6. The industrial oxygen gas (O_2) was injected as oxidant to oxidize U(IV) into U(VI). The ratio of CO_2 and O_2 was fixed at 1:10 with a constant injection pressure of 2 bar throughout the batch process. At the end of each test, solid and liquid phase separation was accomplished by centrifugation. For centrifugation, 15 mL of the batch were transferred to 50 mL polycarbonate centrifuge tubes and centrifuged at 4000 rpm for 15 min. The volume of leachate recovered was measured and recorded. The pH and Eh were monitored with a Hanna Instruments HI 98121 device (Hanna, Hungary). The Hanna meter was calibrated for pH function with 4.0 and 7.0 buffers. The Eh calibration was performed using a Zobell solution.

2.6. Characterization Techniques

Alkalinity of the leachate was measured by titration with HCl. The endpoint of the titration was determined by a color change. Briefly, methyl orange and phenolphthalein were used to detect the bicarbonates and carbonates, respectively. The liquid phase was analyzed for U_3O_8 through fluorimetry and spectrophotometrically using dibenzoyl methan as coloring agent, depending on the concentration of U_3O_8 in the leached solution. The cations Ca^{2+}, Mg^{2+}, Fe^{2+}, Fe^{3+}, Na^+, and K^+ were measured by acid-base titration and flame photometry, respectively. The anions (SO_4^{2-}, Cl^-) were analyzed spectrophotometrically. All liquid samples were filtered using a syringe. The elemental composition and SiO_2 were analyzed through atomic absorption spectrometer and spectrophotometrically, respectively, while the whole rock was analyzed through XRF. For solid material, U assays were carried out by taking 5 g of pulverized sample of the finely milled to 200 mesh size. The 5 g sample was moistened with (1.00 N) HNO_3 for the removal of carbonates with effervescence, then 2 mL $HClO_4$ and 10 mL HF were added; the sample was then heated to dry in a microwave oven-digester. Again, 10 mL HNO_3 and 10 mL HF were added and the sample was heated to dry again. Followed by addition of 50 mL of (1.00 N) HNO_3, and the mixture was leached for one hour, and finally it was diluted to a volume of 100 mL. Inorganic carbon was analyzed through acidification of the finely grounded sample using concentrated HCl followed by back titration with NaOH. Organic carbon (C org.) was analyzed by oxidizing the sample with a $K_2Cr_2O_7$ and H_2SO_4 mixture followed by back titration using freshly prepared ferrous ammonium sulphates. Total carbons were analyzed in this manner and cross-checked using a sulfur analyzer (CS-800 Eltra, Germany).

XRD patterns of the U ore sample were recorded using a diffractometer (Bruker D8 ADVANCE, Karlsruhe, Germany) with CuKα radiation (40 kV, 40 mA) in a continuous scanning mode, and the 2θ scanning ranged from 3° to 80°. EVA software (V4.2.1, developed by Bruker Corporation) was used to analyze the mineral composition with XRD data using the default crystallography open database COD 2013. An EPMA (JXA-8230) was used to identify the U ore mineral. A small sample (<0.5 g) of material was drawn from the fraction that was expected to be enriched with U mineral. The feed ore and leached residue sediment samples were characterized using a scanning electron microscope (SEM) (NovaNanao 450, Fei Czech, Co., Ltd.), while the distribution of elements was detected by energy-dispersive X-ray spectrometry (EDS) (Oxford X-Max^N).

3. Results and Discussion

3.1. Mineral Association and Geochemical Characteristics

Mineral association in the U ore shown by sample TCS-04 is listed in Table 4 and the XRD pattern is shown in Figure 1. It shows that quartz and mica are the predominant phases with percentages accounting for 30.40% and 20.30%, respectively. The percentages of albite and labradorite are 14.40% and 10.80%, respectively. The quartz is angular with a 0.05 to 0.4 cm size suggesting a nearby source. The oligoclase and pyroxene have percentages of 8.0% and 7.7%, respectively, along with orthoclase accounting for 3.40%. Mostly weathered oligoclase shows polycrystalline orthoclase aggregates. The distribution of calcite is very uneven and it can be localized as calcareous cemented masses or calcareous cemented thin layers. The content of calcium carbonate in the non-calcareous uncemented core samples is about 1.50%.

Table 4. Mineral analysis (%) of uranium ore by XRD.

Sample	Hematite	Magnetite	Zircon	Epidote	Staurolite	Kyanite	Amphibole	Calcite
TCS-04	0.1	0.4	0.1	0.1	0.1	0.1	0.7	1.5
Sample	**Mica**	**Orthoclase**	**Pyroxene**	**Oligoclase**	**Labradorite**	**Albite**	**Quartz**	
TCS-04	22.3	3.4	7.7	8	10.8	14.4	30.4	

The minor minerals including amphibole, kyanite, staurolite, epidote, zircon, magnetite, and hematite possess percentages below unity. The content of magnetite and hematite is 0.4% and 0.10%, respectively.

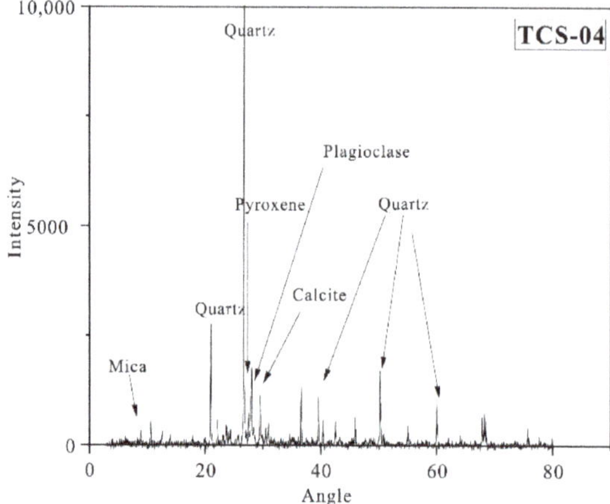

Figure 1. XRD pattern of uranium ore.

The geochemical composition for the four groups' ore samples is shown in Table 5. SiO_2 is the major composition with a percentage range accounting for 57.36–68.87%, followed by Al_2O_3, CaO, and CO_2 with percentage ranges of 10.09–14.89%, 4.75–11.85%, and 4.10–10.35%, respectively. The percentages of oxides of Na, K, Mg and Fe are in the range of 1.00–2.00%, and TiO_2 is less than unity.

Table 5. Geochemical composition of uranium ore samples (wt.%).

Sample	U_3O_8	SiO_2	TiO_2	Al_2O_3	Fe_2O_3	FeO	MnO	CaO	MgO	Na_2O	K_2O	P_2O_5	CO_2
TCS-04	0.0363	64.58	0.32	10.09	1.34	1.36	0.08	7.35	1.66	1.63	1.06	0.09	5.93
TCS-03	0.0540	68.87	0.28	12.78	1.19	1.43	0.05	4.76	4.46	1.82	1.26	0.08	4.31
TCS-02	0.0447	65.78	0.35	14.89	1.77	1.70	0.08	5.88	1.96	1.60	1.01	0.09	4.10
TCS-01	0.0150	57.36	0.21	11.15	0.92	1.26	0.10	11.85	2.72	1.72	0.89	0.08	10.35

The U mineral in the investigated sample (TCS-04) of ore is mainly silica-bearing pitchblende and its composition analysis through EPMA is listed in Table 6. An electronic photograph is shown in Figure 2.

Table 6. Composition of pitchblende (%) analyzed by EPMA.

Sample	EPMA Point	FeO	K_2O	CaO	TiO_2	Al_2O_3	SiO_2	La_2O_3	Ce_2O_3	PbO	Sm_2O_3	Dy_2O_3
TCS-04	3-001	1.30	-	1.98	0.01	7.10	30.93	0.03	0.30	-	-	-
TCS-04	3-003	0.53	-	1.61	0.04	1.06	14.76	-	0.40	0.08	0.06	0.02
TCS-04	3-006	1.29	-	2.18	0.07	2.83	23.78	-	0.46	-	0.04	0.14
TCS-04	3-007	1.12	-	2.20	-	2.52	21.38	0.04	0.44	-	-	0.33
TCS-04	3-008	0.82	-	2.03	0.20	2.41	36.33	0.02	0.15	0.03	0.08	0.06
TCS-04	3-009	1.58	0.09	1.75	-	5.35	26.87	-	0.18	0.02	-	0.04

Sample	EPMA Point	Ho_2O_3	Yb_2O_3	Y_2O_3	MoO_3	Gd_2O_3	UO_2	Na_2O	MgO	P_2O_5	Pr_2O_5	Nd_2O_3
TCS-04	3-001	-	-	0.07	0.05	0.01	46.35	0.98	1.38	0.86	0.86	0.03
TCS-04	3-003	0.03	-	0.19	0.06	-	63.83	2.64	0.17	0.79	0.79	0..216
TCS-04	3-006	0.04	-	0.29	-	0.11	62.64	1.10	0.59	0.91	0.04	0.27
TCS-04	3-007	0.07	0.08	0.81	-	0.18	34.32	0.94	0.48	1.16	0.05	0.33
TCS-04	3-008	-	-	0.05	-	-	45.74	0.54	0.20	0.92	0.03	-
TCS-04	3-009	-	-	-	-	0.07	56.26	1.59	0.83	0.82	-	0.02

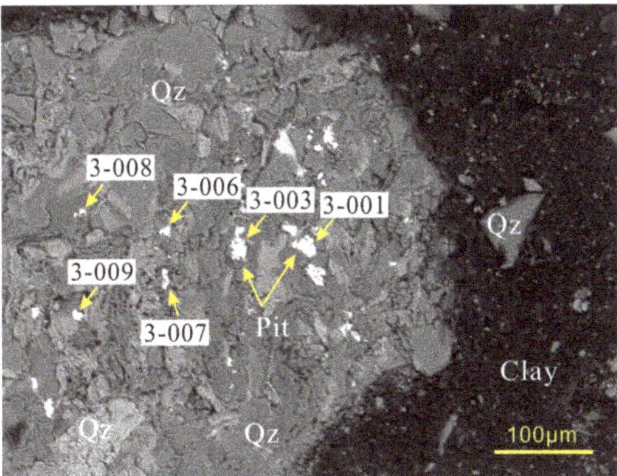

Figure 2. The backscattered electronic photograph of the uranium ore sample. Qz: quartz; Pit: pitchblende.

The U oxide content results of the examined mixed ore sample (TCS-04) and of the ore samples (TCS-03, TCs-02, and TCs-01) with three different grades (high, medium, and low) are shown in Table 7. The geochemical information about sample TCS-04 reflects an overview of the ore deposit with respect to presence of special element characteristics. This sample has an average of 363 ppm of U_3O_8 with a U(IV)/U(VI) ratio of 21 (Table 7) and shows the mobility of U in terms of the U oxidation state. This sample analysis shows that 4.55% of U_3O_8 in the ore deposit may exist in the oxidized form (UO_2^{2+}). The oxidized form may be leachable without injecting an oxidant during the leaching process. However, 95.19% of U_3O_8 in the ore deposit exists in the reduced form (U(IV)), and might be made leachable by injecting an oxidant during the leaching process. A similar composition was observed in the other three groups (TCS-03, TCS-02, and TCS-01), where the oxidized form of U accounted for 12.23%, 9.50%, and 7.08%, respectively. The presence of the U species in ore samples was attributed to the evolution of redox conditions. This is indicated by the Fe species (Fe^{2+} and/or Fe^{3+}) being positively correlated to U(VI), as shown in Figure 3. The general reactivity of Fe significantly influences U geochemistry, and Fe (Fe^{2+} and/or Fe^{3+}) forms mostly oxide, oxyhydroxide, and S-mineral compounds with its substitution in U-minerals commonly [42]. Moreover, there is no well-defined correlation between the U species and organic carbon. It probably suggests that the oxidized U associated with inorganically dominated carbon as compared to organics. Recently, a similar study was documented from the eastern Suliman range of Dear Ghazi Khan, Pakistan [37,38] showing that U ore was mostly associated with organic matter (probably petroleum) as well as other phases of the ore (e.g., biotite). Consequently, the geochemistry of U-containing minerals indicates a relatively pre-complexed environment of the investigated ore deposit, where the occurrence of U(IV) was favored as compared to U(VI). It in turn indicates that the injection of oxidant was needed for the mobilization of the reduced form of U during in situ leaching. A similar case has been reviewed and reported by Shamim Akhtar [43], showing that U deposits in Pakistan were formed in a reduced environment. Various studies have reported that U solubility is affected by its speciation and aqueous chemistry including organic matter (OM) concentrations [44–46]. Low OM concentration was shown to promote U solubility, whereas high OM concentration may immobilize U [47]. Of all the inorganic ligands, carbonates are more active ones to complex U, as compared to others such as SiO_2^{4+} and PO_4^{3-} [48,49].

3.2. Neutral Leaching

Different leaching methods including neutral, acidic, and alkaline have been reported in U mining sites [50–55]. The present study investigated the processing of neutral leaching with the pH value being monitored in the experimental scope. These operational conditions were suggested to favor the application of the neutral leaching since it avoids excessive consumption of reagents (discussed below).

Table 7. Special characteristics of uranium ore.

Sample	U_3O_8 (ppm)	U_{Total} (ppm)	U(VI) (ppm)	U(IV) (ppm)	U(IV) (%)	U(VI) (%)	U(IV)/U(VI)	$C_{org.}$ (%)	$C_{inorg.}$ (%)	Fe^{3+} (%)	Fe^{2+} (%)	SO_3 (%)
TCS-04	363	308	14	293	95.19	4.55	21	0.08	1.62	0.94	1.06	1.06
TCS-03	540	458	56	401	87.57	12.23	7	0.05	1.18	1.28	1.32	1.11
TCS-02	447	379	36	342	90.23	9.50	10	0.06	1.12	1.26	1.32	1.32
TCS-01	150	127	9	117	91.98	7.08	13	0.10	2.82	0.64	0.98	0.98

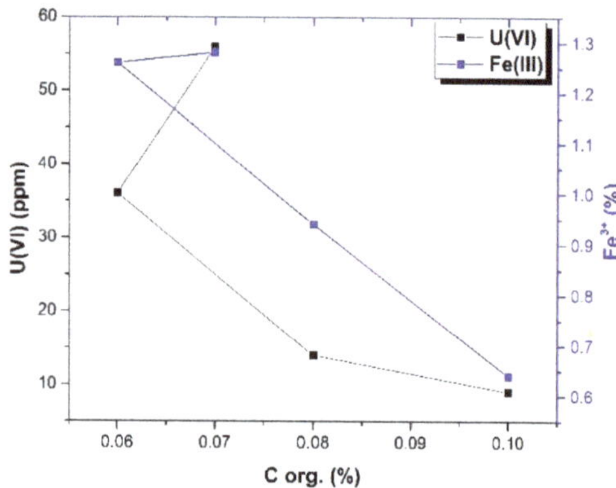

Figure 3. Trend of U(VI) and Fe(III) with organic carbon (C org.) in uranium ore.

3.2.1. Effect of Bicarbonate Concentration and Chemical Dynamics during Leaching

Concentrations of HCO_3^- are shown as a function of time in Figure 4. Overall, two trends were observed during the course of experiments. In the first test, where 0.60 g/L of HCO_3^- concentration was used for the preparation of the lixiviant solution, the U_3O_8 was determined to be 157 mg/L after equilibrium was obtained, while its initial concentration was of 12 mg/L before starting the batch processing (Figure S1). For the two tests (#2 and #3), 0.80 g/L and 1.00 g/L of HCO_3^- concentrations were used, respectively, and the U_3O_8 was determined to be 25 mg/L and 64 mg/L, with an initial concentration of 12 and 5 mg/L before starting the batch processing, respectively. Similarly, in the fourth test (#4), 1.20 g/L of HCO_3^- concentration was used, and the final concentration of U_3O_8 was determined to be 124 mg/L, with its initial concentration of 3 mg/L. The highest U recovery (86%) was found using a 0.60 g/L of HCO_3^- concentration. The U recovery with a concentration of 1.00 g/L of HCO_3^- was 45%, followed by a 0.8 g/L batch where recovery was 13%. The U recovery with 1.20 g/L of HCO_3^- was 68%. The changes in uranium recovery for the tests are shown in Figure 5.

Consequently, the dissolution of the U was favored more easily with 0.6 g/L of HCO_3^-, and only 8.70 days was needed to achieve high recovery leaching time. The lower limit of HCO_3^- concentrations in the aforementioned tests shows a slightly gentle process fluctuation as compared to those on the higher side, and a general improvement in the recovery was observed as the HCO_3^- concentrations increased from 0.8 to 1.2 g/L during the process. The effect of varying concentrations of the leaching agent (HCO_3^-) on U recovery was studied by Elizângela and Ana [56]. Results obtained from the authors showed that 1.00 mol/L of HCO_3^- was the best concentration, under which >95% recovery was obtained in 24 h via alkaline leaching, while a general improvement in the recovery was observed as the HCO_3^- concentrations increased. However, the highest recovery of 86%, as a special result, was obtained from test #1 with 0.6 g/L of HCO_3^-. The lowest recovery of the 0.8 g/L HCO_3^- test may be due to lower bicarbonates and the constantly stable concentration of HCO_3^- as U recovery may highly correlated with it [57]. The geochemical analysis results of ore and residuals are shown in Table 8.

It can be seen that the CaO concentration in all test residues was decreased significantly. This may be due to the injection of CO_2 gas causing the dissolution of calcite (Figure 6), as shown by Equation (1):

$$CaCO_3 + CO_2 + H_2O \rightarrow Ca^{2+} + 2HCO_3^- \quad (1)$$

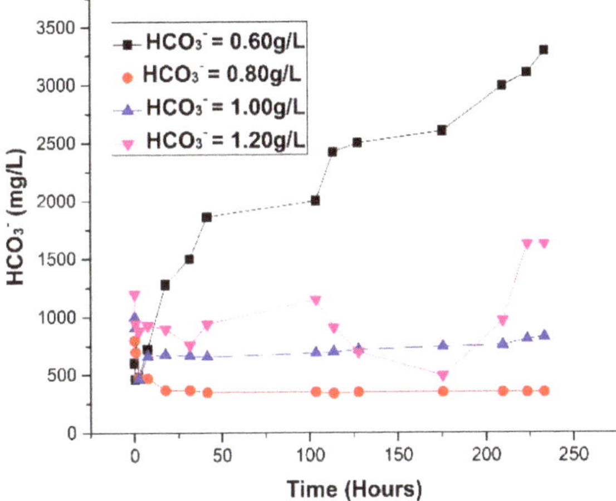

Figure 4. Change in HCO_3^- concentration in the leaching solution.

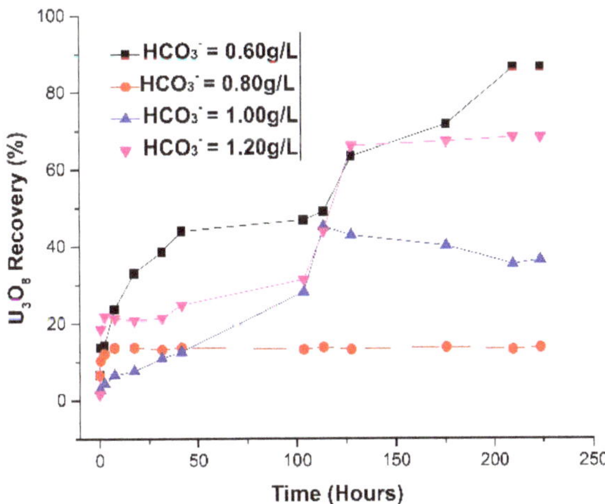

Figure 5. Change in uranium recovery in the leaching solution.

The maximum dissolution of calcium with a concentration of 0.80 g/L of HCO_3^- was 2.33%, followed by the 1.0 g/L batch where the dissolution was 1.56%, while U recoveries were 13% and 45%, respectively. In contrast, the calcium dissolution with 0.60 g/L of HCO_3^- was 1.14%, and in the 1.20 g/L batch, 0.56% calcite was dissolved, with U recoveries being 86% and 68%, respectively. The relationship between U recovery and calcite dissolution is not totally positively correlated. However, 0.60 g/L concentrations of HCO_3^- test have comparatively higher U_3O_8 recovery than that of 1.20 g/L HCO_3^- test with a calcite dissolution range accounting for 1.14% and 0.56%, respectively.

Table 8. Geochemical analysis of ore and residues.

Sample	SiO_2	Al_2O_3	Fe_2O_3	FeO	MnO	CaO	MgO	Na2O	K_2O	P_2O_5	CO_2
TCS-04						%					
Before leaching	64.58	10.09	1.34	1.36	0.08	7.35	1.66	1.63	1.06	0.09	5.93
After leaching (Test #1)	62.16	9.69	1.23	1.67	0.08	6.21	1.34	1.66	1.40	0.08	5.99
After leaching (Test #2)	62.81	10.17	1.10	1.45	0.06	5.02	1.31	1.73	1.28	0.06	5.97
After leaching (Test #3)	64.83	9.84	1.26	1.60	0.08	5.79	1.48	1.83	1.26	0.09	6.21
After leaching (Test #4)	62.42	10.02	1.19	1.65	0.06	6.79	1.34	1.67	1.23	0.08	5.99

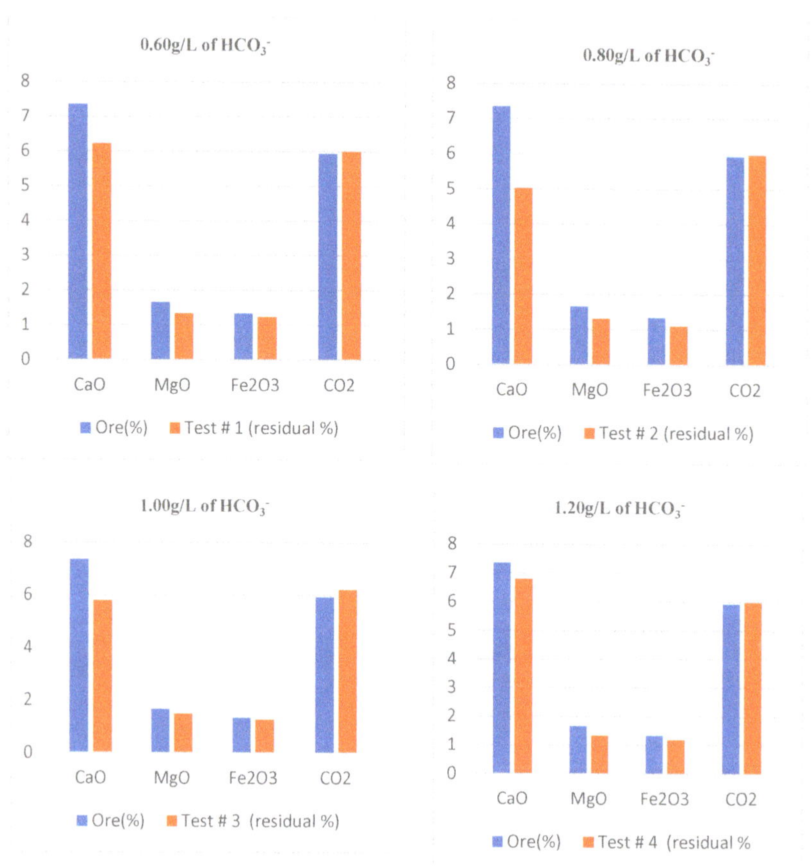

Figure 6. Concentration of oxides before and after leaching.

3.2.2. Effect of pH

The pH decreases quickly and then it remains stable (Figure 7); in addition, its relationship to Eh shows regular patterns (Figure 8). Different patterns of leaching as a function of pH were also observed (Figure 9). The leaching recovery of U increased with changing pH in the early stages of the process and then there was no further change in U recovery. For example, in the test using 0.6 g/L of HCO_3^-, high U recovery was found, while the pH value reduced from 8 to 6.57. Mean and final values recorded for pH for each of the leaching protocols are shown in Table 9. Probably, the plausible mechanism may be due to the injection of CO_2 gas. CO_2 dissolved in water results in reducing pH of the system. The carbonate/bicarbonate ion in turn promotes the release of adsorbed U(VI) from the U-bearing mineral. The released U(VI) dissolved into the liquid phase depends upon contact time and mineral charge density. In addition, increasing H^+ results in the dissolution of gangue minerals like calcite (Figure 6). It can be seen (Figure 9) that the lowest pH value and highest Eh value (most oxidizing conditions) were observed for the best addition of HCO_3^- (0.6 g/L). It should also be noted that pH showed almost a negative correlation with recovery (Figure 9a), i.e., more recovery resulted towards the lower pH of the neutral range, resulting in the formation of more H^+ in the solution. Briganti et al. [28] conducted a study using chemical leaching and adsorption tests with a simple modeling using PHREEQC. They found that the geochemical behavior of U during ignimbrite-water interaction was controlled mainly by temperature, pH, and solution chemistry (especially alkalinity). The main results of their work indicated that U was more easily mobilized by a slightly basic solution (pH 7.5). However, the results obtained from the research revealed that U was more easily mobilized by a slightly acidic solution (pH range 6 to 6.6) during nearly neutral leaching conditions. In many respects, leaching behavior as reflected by pH leaching tests and related characterization provide a better means of assessing environmental impact than an analysis of total composition, such as how the solubility changes if in situ pH changes occur. The pH is one of the key parameters that determines heavy metal mobility depending upon soil and sediment properties.

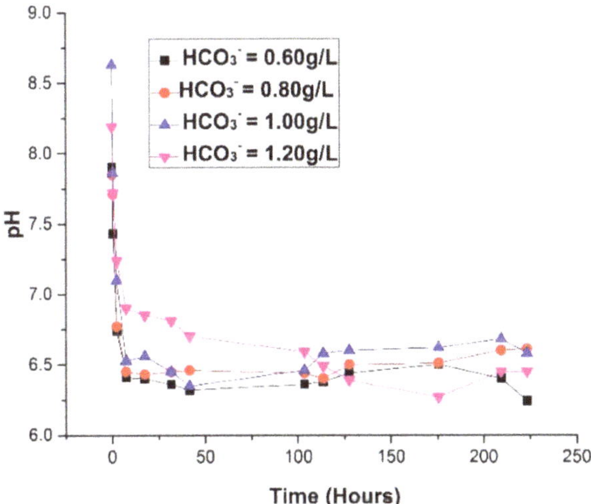

Figure 7. Change in pH measurements in the leaching solution.

The aqueous U(VI) carbonate system has been thoroughly studied by many researchers [58–64]. It is well accepted that the three monomeric complexes of general formula $UO_2(CO_3)$, $UO_2(CO_3)_2^{2-}$, and $UO_2(CO_3)_3^{4-}$ present under the appropriate conditions [65].

In neutral to carbonate media, U is converted into a series of carbonate complexes (UO_2CO_3 (maximum fraction at pH 5), $UO_2(CO_3)_2^{2-}$ (maximum fraction at pH 6.5), and $UO_2(CO_3)_3^{4-}$ (maximum fraction at pH 10 to 11)) when the pH of the solution is increased.

Nevertheless, in all the tests performed, the concentration of CO_3^{2-} was not detected in the leached solution during analysis because of the pH range condition was 6 to 6.6 (which indicates that $UO_2(CO_3)_2^{2-}$ is the main carbonate complex in the condition studied in this work). In such cases, the recovery of U from a nearly neutral medium is possibly due to the formation of stable U carbonate complex-like $UO_2(CO_3)_2^{2-}$ [66] via Equation (2):

$$UO_2 + 1/2O_2 + 2HCO_3^{2} \rightarrow UO_2(CO_3)_2^{2-} + H_2O \tag{2}$$

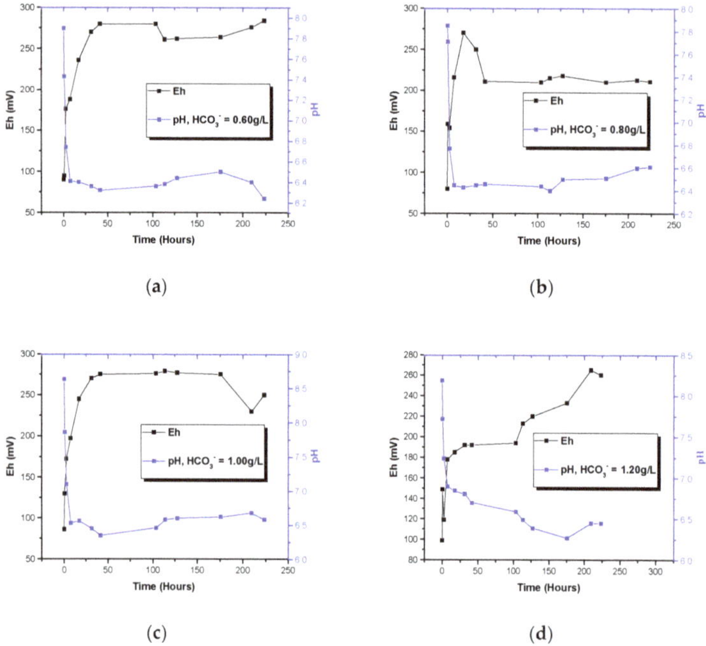

Figure 8. Trend of Eh and pH in leaching solution with different HCO_3^- concentrations: (**a**) 0.60 g/L; (**b**) 0.80 g/L; (**c**) 1.00 g/L; (**d**) 1.20 g/L.

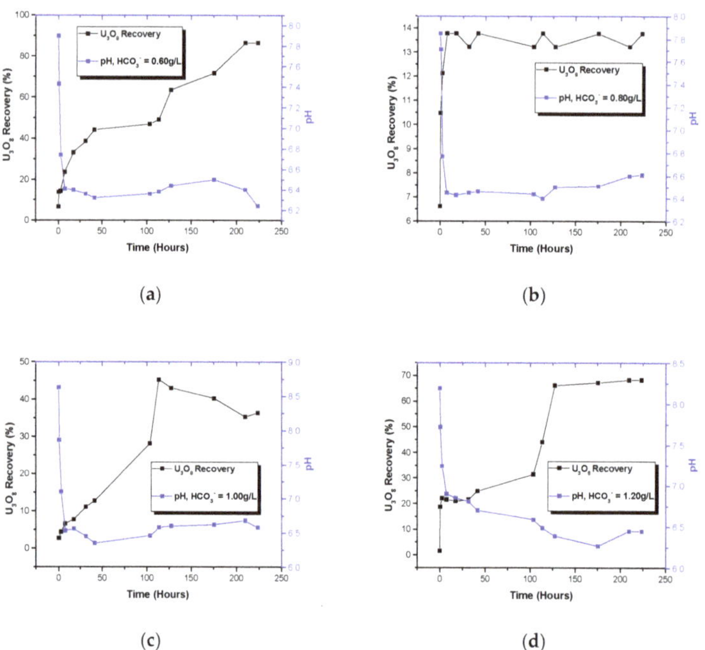

Figure 9. Trend of U_3O_8 recovery and pH in leaching solution with different HCO_3^- concentrations: (**a**) 0.60 g/L; (**b**) 0.80 g/L; (**c**) 1.00 g/L; (**d**) 1.20 g/L).

Table 9. pH, Eh, and recovery changes against each of the leaching protocols investigated.

Sample	HCO_3^-	Leachate pH		Leachate Eh (mV)		U_3O_8 Recovery
	g/L	Mean	Final	Mean	Final	%
TCS-04	0.60	6.60	6.24	228	284	86
do	0.80	6.7	6.61	201	211	13
do	1.00	6.84	6.58	228	250	45
do	1.20	6.85	6.45	192	260	68
TCS-03	0.6	6.65	6.92	209	265	65
TCS-02	do	6.86	6.48	185	225	68

This is a stable U compound in the determined pH range. Many researchers have reported that when pH > 6, U(VI) is dominant in the presence of more complexes such as uranyl hydroxyl complexes and polynuclear uranyl hydroxyl complexes, and the reactivity of Fe inhibits U(VI) [67,68]. It may be the advantage of relatively less Fe (Table 4) in the investigated ores in terms of the experimental scope (pH 6 to 6.6). Under near-neutral conditions, U forms soluble complexes with carbonate and phosphates. Ma et al. [31] conducted an adsorption study of the U species and reported that these species are also influenced by the charge of the mineral surface, depending on the pH of the sorbate solution. Recently, Zhou et al. [46] reported from ISL research from mining sites that, as the pH value decreases, the adsorbed UO_2^{2+} can be replaced by protons and U concentration increases significantly in the leached solution.

3.2.3. Effect of U(VI) Initial Concentration

The effect of different U(VI) initial concentrations (in the solid phase) on U_3O_8 recovery was investigated. The results are listed in Table 10 and shown in Figure 10, showing that recovery was 86% at the lower initial U(VI) concentration (sample TCS-04). A higher U(VI) initial concentration in the solid phase leads to a lower recovery of U_3O_8 (Figure 10). This may be due to the saturation of Fe^{2+} + Fe^{3+} adsorption capacity as its concentration increases with increasing initial U(VI) concentration in the solid phase (Table 10). As the concentration of U(VI) increases in the liquid phase, UO_2^{2+} ions compete for adsorption sites or available functional groups [69]. It is reported that adsorption capacity continues to increase with the rise of initial U(VI) concentration, which may be due to the presence of more U(VI) ions around Fe [70]. Laboratory experiments focusing on U chemical species have been previously done as well for the purpose of researching in situ leaching of U ore, and have reported that the recovery of leaching U is decided jointly by both U content and its activity, according to Ma et al. [31]. Probably, it seems that another mechanism may be due to the presence of different U(VI) initial concentrations in the solid phase, as it is negatively correlated with U_3O_8 recovery (Figure 10).

Table 10. Initial U^{6+} concentration in rock samples and uranium recovery in nearly neutral media.

Sample	Initial Conc. of U^{6+} (ppm)	Fe^{2+} + Fe^{3+} (%)	U_3O_8 Recovery (%)
TCS-04	14	2.00	86.00
TCS-02	36	2.56	69.00
TCS-03	56	2.60	66.00

A similar nature of correlation between U(VI) initial concentration (in the solid phase) and leaching recovery was examined under alkaline leaching. A total of five samples with varying U(VI) initial concentrations were tested. These were the five ore samples (FR1T-860C-28, FR1T-859C-24, FR1T-860C 26, FR1T-860C-20, and FR1T-860C-24, Table 11) out of a total of 36 collected ore samples, which were used in the neutral experimental study after sample grouping. The objective was to confirm the effect of this special influential parameter on leaching recovery comparison as a trial through other leaching media. The results are listed in Table 11 and shown in Figure 10, showing that recovery was 58% at the lower initial U(VI) concentration, while the recovery decreased by increasing U(VI) initial concentration in the solid phase. The parallel leaching experimental study showed a similar negative correlation of U(VI) initial concentration with U_3O_8 recovery in the leached solution.

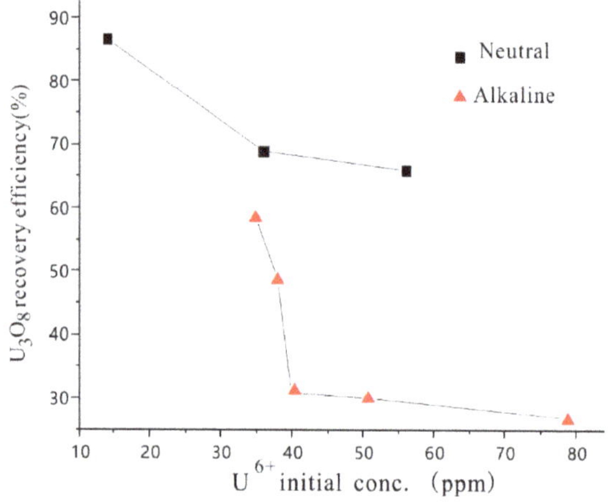

Figure 10. Trend of uranium recovery (in leaching solution) with initial U^{6+} concentration (conc.) in rock samples.

Table 11. Initial U^{6+} concentration in rock samples and uranium recovery in alkaline media.

Sample	U_3O_8 (ppm)	U^{6+} (ppm)	U^{4+} (ppm)	NH_4HCO_3 (g/L)	H_2O_2 (g/L)	pH (Mean/Final)		U_3O_8 Recovery (%)
FR1T-860C-28	390	35	356	2.00	1.00	8.27	8.09	58
FR1T-859C-24	460	38	425	2.00	1.00	8.47	8.08	48
FR1T-860C-26	408	40	305	2.00	0.50	8.38	8.28	31
FR1T-860C-20	500	51	372	2.00	1.00	8.32	8.78	30
FR1T-860C-24	442	79	295	3.00	0.50	8.18	7.79	26

3.2.4. Effect of Redox Conditions

Trends of redox potential (Eh) values are shown as a function of time (Figure 11). Different patterns of oxidation as a function of Eh were observed. The test with 0.6 g/L of HCO_3^- has a high leaching rate, and the Eh value was enhanced more than three times (from 90 to 284 mV), while the test with 1.2 g/L of HCO_3^- enhanced the Eh value less than three times and a similar Eh increment was observed in the tests with 0.80 and 1.0 g/L of HCO_3^- concentrations, as shown in Figure 12. For the most part, the leaching of U increased with increasing Eh, although there was then no further change in U recovery at slightly increased Eh. Mean and final values recorded for Eh for each of the leaching protocols are shown in Table 9. It should also be noted that Eh showed almost a positive correlation with recovery, i.e., a higher recovery rate in higher oxidizing conditions at the later stage of the process (Figure 12), in the test with 0.6 g/L of HCO_3^-. The slope of the figure from different curves was 0.409, 0.050, 0.204, and 0.410 against HCO_3^- concentrations of 0.6, 0.8, 1.0, and 1.2 g/L, respectively. The Eh was almost directly affecting U mobilization. It is well accepted that Eh of the solution is the key factor affecting the dissolution of tetravalent uranium during the leaching of U, which is related to the composition and content of the variable valence ions in the solution. Zhou et al. [46] reported that change in the uranium concentration in the leached solution is time-lagged and its peak occurred synchronously with that of high Eh.

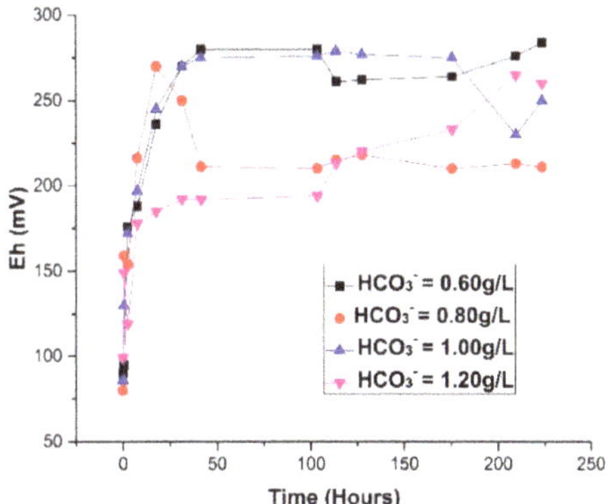

Figure 11. Change in Eh measurements in the leaching solution with different HCO_3^- concentrations.

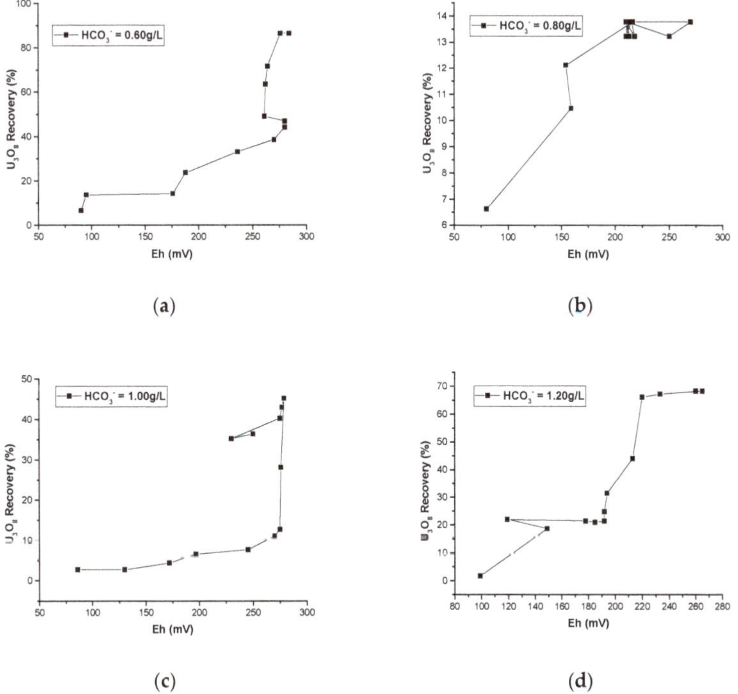

Figure 12. Uranium recovery and Eh variations in leaching solution with different HCO_3^- concentrations: (**a**) 0.60 g/L; (**b**) 0.80 g/L; (**c**) 1.00 g/L; (**d**) 1.20 g/L.

4. Conclusions

This investigation included a geochemical characterization study and nearly neutral (pH: 6–6.6) leaching mechanisms of uranium (U) ore from the ELZPA deposit. The ore mineral was determined to be dominated by pitchblende, as revealed by EPMA and SEM-EDS analytical results. Leaching experiments with a $CO_2 + O_2$ system were performed to investigate U mobility and transformation. Parameters of pH, Eh, alkalinity, and major ion(s) were shown to influence U mobilization. U_3O_8 recovery was achieved to a high level (86%), mainly in the form of a stable $UO_2(CO_3)_2^{2-}$ complex in the leached solution. The plausible U migration mechanism may be due to the oxidizing conditions, and the dissolved CO_2 gas via a formation of protons releasing oxidized U form the uranium-bearing minerals. However, the injection of CO_2 may cause a dissolution of gangue minerals (i.e., calcite). Comparatively, the overall uranium recovery in different ore group tests was significantly negatively correlated with prevalent organic complexed oxidized U in the solid phase under nearly neutral and alkaline leaching environments. Therefore, further studies are needed to explore such influential parameters within the research scope.

Supplementary Materials: The following are available online at http://www.mdpi.com/2075-4701/10/12/1616/s1, Figure S1: Change in uranium concentration in the leaching solution.

Author Contributions: Conceptualization, Z.S. and Y.Z.; methodology, G.L.; software, F.A.; validation, Y.Z., G.L., G.C. and H.L.; formal analysis, F.A. and K.Z.; investigation, Z.S. and F.A.; resources, Z.S.; data curation, F.A. and K.Z.; writing—original draft preparation, F.A.; writing—review and editing, Y.Z. and G.C.; visualization, F.A.; supervision, Z.S.; project administration, Z.S.; funding acquisition, Z.S. All authors have read and agreed to the published version of the manuscript.

Funding: This research was funded by the program of the National Natural Science Foundation of China (NSFC) (Nos. 41772266, 42072285 and 41662015). The APC was funded by East China University of Technology.

Acknowledgments: We thank the reviewers for their comments, which have played a great role in improving the manuscript.

Conflicts of Interest: The authors declare no conflict of interest.

References

1. IAEA, 20094. International Atomic Energy Agency. 2012. The Database on Nuclear Power Reactors. World Nuclear Association 2012. Available online: http://www.iaea.org/pris/ (accessed on 12 November 2012).
2. *Uranium Mining, Processing and Nuclear Energy-Opportunites for Australia*; Department of Prime Minister and Cabinet: Fitzroy Crossing, Australia, 2006; ISBN 0-9803115-0-0 978-0-9803115-0-1.
3. IAEA. *Organization for Economic and Co-Operation and Development (OECD) Nuclear Energy Agency (NEA) and the International Atomic Energy Agency*; IAEA: Vienna, Austria, 2012; p. 489.
4. IAEA, Vienna, IAEA TECDOC 1629. *World Distribution of Uranium Deposits (UDEPO) with Uranium Deposit Classification*; International Atomic Energy Agency: Vienna, Austria, 2009; ISBN 978-92-0-110509-7.
5. Evceeva, L.S.; Fomina, N.L. *Oxidation-Reduction Character Sedimentary Uranium Bearing Rock*; Atomic Energy Press: Beijing, China, 1965; pp. 5–12.
6. Langmuir, D.; Chatham, J.R. Groundwater prospecting for sandstone-type uranium deposits: A preliminary comparison of the merits of mineral-solution equilibria, and single-element tracer methods. *J. Geochem. Explor.* **1980**, *13*, 201–219. [CrossRef]
7. Dannev, F.Y.; Strelianov, N.P.; Yao, Z.K. Some geochemical characteristics of uranium in the external ore forming process. *Earth Environ.* **1982**, *6*, 45–47.
8. Morrison, S.J.; Tripathi, V.S.; Spangler, R.R. Coupled reaction/transport modeling of a chemical barrier for controlling uranium (VI) contamination in groundwater. *J. Contam. Hydrol.* **1995**, *17*, 347–363. [CrossRef]
9. Fuller, C.C.; Bargar, J.R.; Davis, J.A.; Piana, M.J. Mechanisms of uranium interactions with hydroxyapatite: implications for groundwater remediation. *Environ. Sci. Technol.* **2002**, *36*, 158–165. [CrossRef] [PubMed]
10. Noubactep, C.; Schöner, A.; Meinrath, G. Mechanism of uranium removal from the aqueous solution by elemental iron. *J. Hazard. Mater.* **2006**, *132*, 202–212. [CrossRef]
11. Yang, X.Y.; Ling, M.X.; Sun, W.; Liu, C.Y. Study on the ore-forming condition and occurrence of uranium minerals in sandstone-type uranium deposits from Ordos basin, Northwest China. *Geochim. Cosmochim. Acta* **2006**, *70*, A720. [CrossRef]
12. Yue, S.; Wang, G. Relationship between the hydrogeochemical environment and sandstone-type uranium mineralization in the Ili basin, China. *Appl. Geochem.* **2011**, *26*, 133–139. [CrossRef]
13. Alam, M.S.; Cheng, T. Uranium release from sediment to groundwater: influence of water chemistry and insights into release mechanisms. *J. Contam. Hydrol.* **2014**, *164*, 72–87. [CrossRef]
14. Golubev, V.N.; Dubinina, E.O.; Chernyshev, I.V. Behavior of isotope (18O/ 16O, 234 U/238U) systems during the formation of uranium deposits of the "sandstone" type. *Dokl. Earth Sci.* **2016**, *466*, 28–31. [CrossRef]

15. Tsarev, S.; Collins, R.; Ilton, E.S.; Fahy, A.; Waite, T.D. The short-term reduction of uranium by nanoscale zero-valent iron (nZVI): Role of oxide shell, reduction mechanism and the formation of U(V)-carbonate phases. *Environ. Sci. Nano* **2017**, *4*, 1304–1313. [CrossRef]
16. Lammers, L.N.; Rasmussen, H.; Adilman, D.; Delemos, J.L.; Zeeb, P.; Larson, D.G. Groundwater uranium stabilization by a metastable hydroxyapatite. *Appl. Geochem.* **2017**, *84*, 105–113. [CrossRef]
17. Post, V.; Vassolo, S.I.; Tiberghien, C.; Baranyikwa, D.; Miburo, D. Weathering and evaporation controls on dissolved uranium concentrations in groundwater—A case study from northern Burundi. *Sci. Total Environ.* **2017**, *607–608*, 281–293. [CrossRef] [PubMed]
18. Maher, K.; Bargar, R.; Brown, E. Environmental speciation of actinides. *Inorg. Chem.* **2012**, *52*, 3510–3532. [CrossRef] [PubMed]
19. Murakami, T.; Ohnuki, T.; Isobe, H.; Sato, T. Mobility of uranium during weathering. *Am. Mineral.* **1997**, *82*, 888–899. [CrossRef]
20. Screenivas, T.; Rao, K.A.; Rao, M.M.; Rajan, K.C.; Serajuddine, M.; Karthikayini, P.; Padmanabhan, H.N.P. *Autoclave Leaching of Refractory Uranium Minerals*; Allied Publishers: New Delhi, India, 2010.
21. Gilligan, R.; Nikoloski, A.N. The extraction of uranium from brannrite. A literature review. *Miner Eng.* **2015**, *71*, 34–48. [CrossRef]
22. Simon, R.B.; Thiry, M.; Schmitt, J.M.; Lagneau, V.; Langlais, V.; Bélières, M. Kinetic reactive transport modelling of column tests for uranium In Situ Recovery (ISR) mining. *Appl. Geochem.* **2014**, *51*, 116–129. [CrossRef]
23. Erik, H. On the Leaching Behavior of Uranium-Bearing Resources in Carbonate-Bicarbonate Solution by Gaseous Oxidants. Ph.D. Thesis, Colorado School of Mines, Golden, CO, USA, 2013.
24. Kacham, A.R.; Suri, A.K. 2013 Application of a topochemical reaction model to predict leaching behavior of high carbonate uranium ores in alkaline solutions: An experimental case study. *Hydrometallurgy* **2014**, *141*, 67–75. [CrossRef]
25. Chris, M.; Ben Schiffer, P.G.; Jack Fritz, P.G. Low pH uranium ISR: historical projects, current practices and pathway to future use in Wyoming. In *A White Paper Prepared for: Prepared for Strata Energy*; Strata Energy, Inc.: Oshoto, WY, USA, 2017.
26. Valérie, C.; Rudy, S. The use of Leaching tests to study the potential mobilization of heavy metals from soils and sediments: A Comparison. *Water Air Soil Pollute* **2008**, *191*, 95–111.
27. Dason, G.K.W.; Sidiq, H.; Pearce, J.; Gao, J.F.; Golding, S.D.; Rudolph, V.; Li, Q.; Xing, H. *Review of Laboratory-Scale Geochemical and Geomechanical Experiments Simulating Geosequestration of CO_2 in Sandstone, and Associated Modelling Studies*; ANLEC Project 3-1110-0101; Australian National Low Emissions Coal Research and Development: Manuka, Australia, 2013.
28. Briganti, A.; Armiento, G.; Nardi, E.; Proposito, M.; Tuccimei, P. Understanding uranium behavior in a natural rock-water system: leaching and adsorption tests on the Tufo Rosso a Scorie Nere ignimbrite. *Environ. Earth Sci.* **2017**, *76*, 680. [CrossRef]
29. Xu, L.; Yang, H.; Zhou, Y.; Li, G. Neutral leaching of uranium ore from a sandstone-type deposit, Nonferrous Metals. *Extr. Metall.* **2020**, *3*, 38–44.
30. Zhou, Y.; Li, G.; Xu, L.; Sun, Z.; Zhou, Z.; Li, H. Flow field simulation and hydrochemical verification from $CO_2 + O_2$ in-situ leaching of sandstone type uranium deposit, Nonferrous Metals. *Extr. Metall.* **2019**, *1*, 42–47.
31. Ma, Q.; Feng, Z.; Liu, P.; Lin, X.; Li, Z.; Chen, M. Uranium speciation and in situ leaching of a sandstone-type deposit from China. *J. Radioanal. Nucl. Chem.* **2016**, *311*, 2129–2134. [CrossRef]
32. Raza, H.A.; Ahmed, R.; Ali, S.M.; Ahmad, J. Petroleum prospects: Sulaiman sub-basin, Pakistan: Pakistan. *J. Hydrocarb. Res.* **1989**, *1*, 21–26.
33. Hunting Survey Corporation. *Reconnaissance Geology of Part of West Pakistan: A Colombo Plan Cooperative Project*; Hunting Survey Corporation: Toronto, ON, Canada, 1960.
34. Baker, M.A.; Jackson, R.O. *Geological Map of Pakistan 1:2,000,000: Geological Survey of Pakistan*; Fahad Publishing Company: Balochistan, Pakistan, 1964.
35. Kazmi, H.A.; Jan, Q.M. *Geology and Tectonics of Pakistan (Handbook)*; Graphic Publishers: Karachi, Pakistan, 1997; ISBN 969-8375-00-7.
36. Shah, S.M.I. *Stratigraphy of Pakistan: Memoirs of Geological Survey of Pakistan*; Ministry of Petroleum and Natural Resources: Islamabad, Pakistan, 2009.

37. Rahman, U.; Nie, F.; Zhang, X.; Zhang, C.; Asim, A.; Zhang, P. Uranium Traps in Phreatic Sandstone-type Prospect, Taunsa Area, Dera Ghazi Khan, Eastern Sulaiman Range, Pakistan: Evidences from Autoradiography and Optical Microscopy. *Int. J. Econ. Environ. Geol.* **2020**, *11*, 65–71.
38. Rahman, U.; Nie, F.; Zhang, X.; Zhang, C.; Saqib, I.; Idrees, S.; Asim, A. Occurrence of a Likely Tuff Bed between the Middle and Upper Siwaliks, Taunsa area, Dera Ghazi Khan, Eastern Sulaiman Range, Pakistan. *Int. J. Econ. Environ. Geol.* **2020**, *11*, 24–34.
39. Pendowski, J.J. *An Assessment of Laboratory Leaching Tests for Predicting the Impacts of Fill Material on Ground Water and Surface Water Quality*; Washington State Department of Ecology: Olympia, WA, USA, 2003.
40. ASTM Standards 2001. *Standard Test Method for Distribution Ratios by the Short-Term Batch Method*; ASTM D4319-93 (Reapproved 2001); ASTM International: West Conshohocken, PA, USA, 2001.
41. Herreweghe, V. Chemical associations of heavy metals and metalloids in contaminated soils near former ore treatment plants: a differentiated approach with emphasis on pHstat-leaching. *J. Geochem. Explor.* **2002**, *76*, 113–138. [CrossRef]
42. Bruggeman, C.; Maes, N. Uptake of uranium (VI) by pyrite under boom clay conditions: Influence of dissolved organic carbon. *Environ. Sci. Technol.* **2010**, *44*, 4210–4216. [CrossRef]
43. Shamim, A.; Yang, X.; Wang, F.Y. Uranium deposits and resources potential in Pakistan: A review. *Sci. Int.* **2015**, *27*, 1293–1296.
44. Manaka, M.; Seki, Y.; Okuzawa, K.; Watanabe, Y. Uranium sorption onto natural sediments within a small stream in central Japan. *Limnology* **2008**, *9*, 173–183. [CrossRef]
45. Luo, W.; Gu, B. Dissolution and mobilization of uranium in a reduced sediment by natural humic substances under anaerobic conditions. *Environ. Sci. Technol.* **2008**, *43*, 152–156. [CrossRef]
46. Tinnacher, R.M.; Nico, P.S.; Davis, J.A.; Honeyman, B.D. Effects of fulvic acid on uranium(VI) sorption kinetics. *Environ. Sci. Technol.* **2013**, *47*, 6214–6222. [CrossRef]
47. Campbell, K.M.; Kukkadapu, R.K.; Qafoku, N.P.; Peacock, A.D.; Lesher, E.; Williams, K.H.; Bargar, J.R.; Wilkins, M.J.; Figueroa, L.; Ranville, J.; et al. Geochemical, mineralogical and microbiological characteristics of sediment from a naturally reduced zone in a uranium-contaminated aquifer. *Appl. Geochem.* **2012**, *27*, 1499–1511. [CrossRef]
48. Guo, Z.; Li, Y.; Wu, W. Sorption of U(VI) on goethite: Effects of pH, ionic strength, phosphate, carbonate and fulvic acid. *Appl. Radiat. Isot.* **2009**, *67*, 996–1000. [CrossRef] [PubMed]
49. De Windt, L.; Burnol, A.; Montarnal, P.; van der Lee, J. Intercomparison of reactive transport models applied to UO_2 oxidative dissolution and uranium migration. *J. Contam. Hydrol.* **2003**, *61*, 303–312. [CrossRef]
50. Su, X.; Han, Q. CO_2 + O_2 in-situ leaching of uranium [G]. In *Collection of Scientific and Technological Papers of Beijing Institute of Chemical and Metallurgical Research of Nuclear Industry*; Beijing Institute of Chemical Metallurgy Nuclear Industry: Beijing, China, 2008.
51. Zhou, Y.; Li, G.; Xu, L.; Liu, J.; Sun, Z.; Shi, W. Uranium Recovery from Sandstone-Type Uranium Deposit by Acid In-Situ Leaching-An Example from the Kujieertai. *Hydrometallurgy* **2020**, *191*, 105209. [CrossRef]
52. Suzuki, S.; Hirono, S.; Awakura, Y.; Majima, H. Solubility of uranous sulfate in aqueous sulfuric acid5olution. *Metall. Trans. B* **1990**, *21*, 839–844. [CrossRef]
53. Gao, B.; Shi, W.J.; Wang, G.H. Study of migration characteristics of solute (uranium) during in-situ leach process. *Uranium Geol.* **2003**, *19*, 100–105.
54. Shi, W.J.; Gao, B.; Wang, G.H. Mechanism of solute migration during in-situ leaching l sandstone type uranium deposit. *East. China Inst. Technol.* **2004**, *27*, 24–32.
55. Yousef, G.; Marcelo, R.M. Leaching behavior and the solution consumption of uranium-vanadium ore in alkali carbonate-bicarbonate column leaching. *Hydrometallurgy* **2016**, *161*, 127–137.
56. Elizângela, A.S.; Ana, C.; Ladeira, Q. Leaching of uranium from the osamu utsumi mine wastes, inb caldas, minas gerais. In Proceedings of the 2009 International Nuclear Atlantic Conference-INAC, Rio de Janeiro, RJ, Brazil, 27 September–2 October 2009; ISBN 978-85-99141-03-8.
57. Yang, X.-Y.; Chen, G.; Zhen, Y.-Z.-G.; MA, S.-R.; Mula, D.-J.; Din, Y.-Q. Discussion on relationship between uranium leaching and bicarbonate concentration by Co_2 + O_2 in situ leaching sandstone type uranium deposit. *Uranium Min. Metall.* **2020**, *39*, 1.
58. Kumar, A.; Rout, S.; Narayanan, U.; Mishra, M.K.; Tripathi, R.M.; Singh, J.; Kumar, S.; Kushuaha, H. Geochemical modelling of uranium speciation in the subsurface aquatic environment of Punjab state in India. *Geol. Min. Res.* **2011**, *3*, 137–145.

59. Gorman-Lewis, D.; Burns, P.C.; Fein, J.B. Review of uranyl mineral solubility measurements. *Chem. Thermodyn.* **2008**, *40*, 335–349. [CrossRef]
60. Choppin, G.R. Actinide speciation in the environment, Radioanal. *Nucl. Chem.* **2007**, *273*, 695–709. [CrossRef]
61. Morss, L.R.; Edelstein, N.M.; Fuger, J.E. *The Chemistry of the Actinide and Transactinide Elements*; Springer: Dordrecht, The Netherlands, 2006.
62. Sutton, M. Uranium Solubility, Speciation and Complexation at High pH. Ph.D. Thesis, Loughborough University, Loughborough, UK, 1999.
63. Newton, T.W.; Sullivan, J.C. *Actinide Carbonate Complexes in Aqueous Solutions: Handbook on the Physics and Chemistry of the Actinides*; Elsevier: Amsterdam, The Netherlands, 1985.
64. Allard, B. *Actinide Solution Equilibria and Solubilities in Geologic Systems*; K.B.S. Technical Report Chalmers University of Technology: Goteborg, Sweden, 1983; pp. 83–85.
65. Song, W.C.; Shao, D.D.; Songsheng, L.U.; Wang, X.K. Simultaneous removal of uranium and humic acid by cyclodextrin modified graphene oxide nanosheets. *Sci. China Chem.* **2014**, *57*, 1291–1299. [CrossRef]
66. Korichi, S.; Bensmaili, A. Sorption of uranium (VI) on homoionic sodium smectite experimental study and surface complexation modeling. *J. Hazard. Mater.* **2009**, *169*, 780–793. [CrossRef]
67. Zhao, D.; Wang, X.; Yang, S.; Guo, Z.; Sheng, G. Impact of water quality parameters on the sorption of U(VI) onto hematite. *J. Environ. Radioact.* **2012**, *103*, 20–29. [CrossRef]
68. Yousef, L.A.; Morsy, A.M.A.; Hagag, M.S. Uranium ions adsorption from acid leach liquor using acid cured phosphate rock: kinetic, equilibrium, and thermodynamic studies. *Sep. Sci. Technol.* **2020**, *55*, 648–657. [CrossRef]
69. Feng, J.; Liang, C.; Wang, L.; Zhang, X. Kinetics of Cr(VI) removal from aqueous solution with nanoscale zero-valent iron. *Sci. Technol. Rev.* **2011**, *29*, 37–41.
70. Dong, W.; Brooks, S.C. Determination of the formation constants of ternary complexes of uranyl and carbonate with alkaline earth metals (Mg^{2+}, Ca^{2+}, Sr^{2+}, and Ba^{2+}) using anion exchange method. *Environ. Sci. Technol.* **2006**, *40*, 4689–4695. [CrossRef]

Publisher's Note: MDPI stays neutral with regard to jurisdictional claims in published maps and institutional affiliations.

Article

Uranium Removal from Groundwater and Wastewater Using Clay-Supported Nanoscale Zero-Valent Iron

Borys Kornilovych [1,2,*], Iryna Kovalchuk [2], Viktoriia Tobilko [1] and Stefano Ubaldini [3]

1 National Technical University of Ukraine "Igor Sikorsky Kyiv Polytechnic Institute", 37 Peremogy Av., 03056 Kyiv, Ukraine; vtobilko@gmail.com

2 Institute for Sorption and Problems of Endoecology, National Academy of Science of Ukraine, 13 General Naumov Str., 03164 Kyiv, Ukraine; kowalchukiryna@gmail.com

3 Istituto di Geologia Ambientale e Geoingegneria, CNR, Area della Ricerca di Roma RM 1—Montelibretti-Via Salaria Km 29,300—Monterotondo Stazione, 00015 Roma, Italy; stefano.ubaldini@igag.cnr.it

* Correspondence: b.kornilovych@gmail.com; Tel.: +38-050-8407970

Received: 17 September 2020; Accepted: 23 October 2020; Published: 26 October 2020

Abstract: The peculiarities of sorption removal of uranium (VI) compounds from the surface and mineralized groundwater using clay-supported nanoscale zero-valent iron (nZVI) composite materials are studied. Representatives of the main structural types of clay minerals are taken as clays: kaolinite (Kt), montmorillonite (MMT) and palygorskite (Pg). It was found that the obtained samples of composite sorbents have much better sorption properties for the removal of uranium from surface and mineralized waters compared to natural clays and nZVI.It is shown that in mineralized waters uranium (VI) is mainly in anionic form, namely in the form of carbonate complexes, which are practically not extracted by pure clays. According to the efficiency of removal of uranium compounds from surface and mineralized waters, composite sorbents form a sequence: montmorillonite-nZVI > palygorskite-nZVI > kaolinite-nZVI, which corresponds to a decrease in the specific surface area of the pristine clay minerals.

Keywords: uranium; clays; nanoscale zero-valent iron; groundwater; wastewater

1. Introduction

Pollution of the water basin in sites where uranium ore is mined and processed is one of the most important environmental problems, requiring an effective and rational solution [1]. The sources of surface and groundwater pollution are, first of all, the areas of development of uranium deposits, storages of wastes from hydrometallurgical processing of uranium ores, sites where uranium where enriched [2–4]. Water pollution with uranium compounds is often accompanied by pollution of other toxic elements, for example, arsenic [5].

Contaminated groundwater in places of extraction and processing of uranium ores, in addition to the high content of compounds, is characterized by high mineralization. The latter is a few grams per litre, mainly due to sulfates of Ca and Mg, which are formed due to the use of sulfuric acid in the technological processes of leaching [6]. In such waters, it is mainly in the composition of negatively charged sulfate or carbonate complexes, which significantly complicates the processes of water purification [7].

Sorption methods using ion exchange resins and synthetic inorganic sorbents are most widely used for deep purification of sewage and surface waters from uranium complexes [8–10]. At the same time, if it is necessary to treat large amounts of treated water, the economic factor becomes principal.

Some of the new effective environmental protection technologies that havebeen developed and widely tested in recent years are technologies based on the use of highly reactive nanoscale zero-valent iron. The use of coarsely dispersed zero-valent iron is quite common in environmental practice in the construction of permeable reaction barriers that are installed in the path of contaminated groundwater and serve as a collector of various toxicants [11,12]. The advantage of using nZVI is the possibility of its use for groundwater purification without the application of permeable reaction barriers, the construction of which requires significant capital costs.

Suspensions based on nZVI can be pumped into the ground along the path of groundwater directly in front of contaminated areas. Further, they are carried with underground streams, promoting the decomposition of organic toxicants or sorption of inorganic pollutants on themselves.

The first successful results on the use of nZVI were obtained during the decomposition of organic toxicants (chlorinated solvents, pesticides, dyes) [13,14]. Subsequently, the effectiveness of this technology was shown for a variety of inorganic pollutants, including heavy metals [15] and natural radionuclides [16,17]. However, the potential risks for natural ecosystems when using nZVI is insufficiently studied so far [18].

The problem with the use of nZVI for in situ and ex situ technologies is the insufficient colloidal stability of its suspensions, easy aggregation, and difficulty in separating nZVI from the treated solution. To solve it, various polymers and surfactants are commonly used [19,20]. Another approach to increase the stability of suspensions is their anchoring onto a solid matrix, which allows not only to increase their stability and stability to oxidation, but also to expand the scope of their application in traditional sorption technologies [21]. Active carbons [22], amorphous silica [23,24], layered double hydroxides [25], carbonized fungi [26], graphene and composites based on it [27,28] and others were successfully investigated.

Some of the most widely used matrices as the supports of nZVIare clay minerals, which have a significant specific surface area and rather high sorption properties in relation to radionuclides and heavy metal ions [29,30]. Iron-containing composite sorbents based on clays have enhanced sorption characteristics in relation to heavy metals and radionuclides compared to the initial minerals [31]. Representatives of almost all major classes of clay minerals were studied as matrixes: kaolinite [32], montmorillonite [33,34], illite [35], palygorskite [36].

While considerable experimental material accumulated on the purification of uranium-containing water using nZVI, there was no comparative study of the sorption properties of iron-containing composites of different composition based on clay minerals. Taking into account the importance of environmental studies based on natural minerals, the removal of uranium by nZVI supported on kaolinite, montmorillonite and palygorskite was investigated, including the removal efficiency of uranium from contaminated groundwater with low and high mineralization.

2. Materials and Methods

The object of the study was a layer silicate kaolinite ($Al_4Si_4O_{10}(OH)_8$) with a structure of 1:1 type. This clay was taken from the Glukhovets deposit (Ukraine), and has the most perfect crystal structure among the kaolinites from other numerous kaolin deposits of Ukraine. Among the representatives of layered silicates with a 2:1 structure (smectites), montmorillonite from the Cherkassk deposit (Ukraine) with a structural formula $(Ca_{0.12}Na_{0.03}K_{0.03})_{0.18}(Al_{1.39}Mg_{0.13}Fe_{0.44})_{1.96}(Si_{3.88}Al_{0.12})_{4.0}O_{10}(OH)_2 \times nH_2O$ was chosen. Fibrous clay mineral palygorskite (also known as attapulgite) with a structural formula $Mg_5Si_8O_{20}(OH)_2 \times 4H_2O$ was also taken from the Cherkassk deposit (Ukraine).

The purification of natural minerals from impurities of quartz, feldspars, carbonates, aluminium and iron oxides was carried out.X-ray powder diffraction (XRD) patterns were recorded on X-ray diffractometer DRON-4-07 (Russia) in the range of 2–40° (2θ) using CuKα irradiation.

The parameters of the porous structure: specific surface area (SSA), total pore volume (V), average pore radius (r) of the natural and composite sorbents were determined by the BET method from nitrogen adsorption isotherms obtained on a Nova 2200e gas analyzer (USA), pore distribution(Rmax)—by BJH and DFT methods [37].

To obtain composite sorbents "clay mineral/nZVI" with a mass ratio nZVI: clay mineral~0.2: 1 modified method was used [38,39]. To do this, in a 0.2M solution of $Fe(NO_3)_3 \cdot 9H_2O$, weighed clay was introduced.The resulting suspension (pH2) was quantitatively transferred to a three-necked flask and the ion Fe^{3+} reduction process was performed with a solution of sodium borohydride ($NaBH_4$) in a nitrogen atmosphere. The resulting composite sorbent was then separated from the liquid phase by centrifugation and washed three times with alcohol. The resulting precipitate was dried under vacuum at a temperature of 60 °C and crushed to obtain a fraction ≤0.1 mm.

Water purification from uranium ions was performed using natural clay minerals and composite sorbents. Solutions were prepared with distilled water and uranyl trihydrosulfate salt ($UO_2SO_4{\cdot}3H_2O$). 1M NaCl solution to create ionic strength (~0.01) was used. The pH of the solutions was adjusted with 0.1M solutions NaOH and HCl.

Mineralized waters were used solutions, whose composition corresponded to the composition of underground mineralized water near the liquid waste storage facility for hydrometallurgical processing uranium ores at the Centre of Ukrainian Uranium Industry (Zhovti Vody) by anions: HCO_3^-—450; Cl^-—180; SO_4^{2-}—2830; NO_3^-—130 mg/l and by cations: Ca^{2+}—576; Mg^{2+}—209; ($Na^+ + K^+$)—391; NH_4^+—0.92; Ni^{2+}—<0.05; Cu^{2+}—<0.03; Co^{2+}—<0.06; Mn_{sum}—0.10; Zn^{2+}—<0.01; Pb^{2+}—<0.19; Cd^{2+}—<0.01; Fe_{sum}—0.05 mg/L [40]. The output solutions were prepared on the basis of the corresponding sodium salts, the total salt content was—5280 mg/L, pH 7.2.

Sorption of uranium ions was performed under static conditions at room temperature with continuous shaking of the samples for 1 h (the volume of the aqueous phase—50 mL, the amount of sorbent—0.1 g). Inthe end, the liquid phase was separated by centrifugation (6000 rpm) and the equilibrium metal concentration was determined spectrophotometrically (UNICO 2100UV) using an Arsenazo III reagent at a wavelength of 665 nm.

3. Results

X-ray analysis of samples shows the presence of only small impurities in pristine minerals (Figure 1).

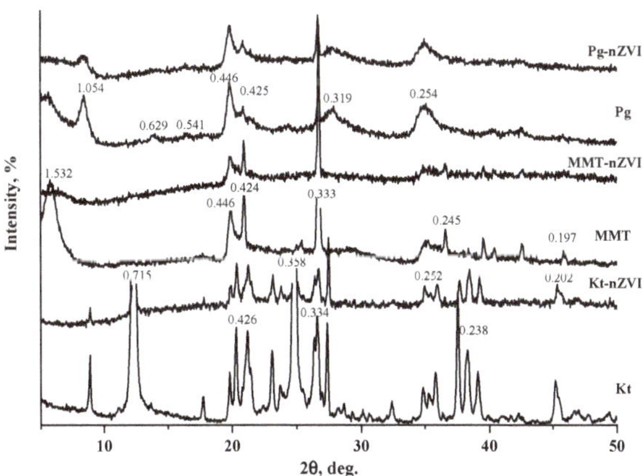

Figure 1. X-ray powder diffraction (XRD) patterns of pristine and nanoscale zero-valent iron (nZVI)—modified clay minerals.

On diffractograms (XRD) of all modified minerals, there are weak reflections at 0.252 and 0.202 nm corresponding to crystalline phases of zero-valent iron (α-Fe), iron oxide (FeO), and also, in time, smaller quantities of goethite (FeOOH). Clays kept their structures in mixtures, since no structural changes between neat and supported clays were found.The amount of iron applied to the surface is, according to the chemical analysis of the solution after treatment of the modified samples with hydrochloric acid, 0.17 mg/g for all minerals.

By nature, the nitrogen sorption isotherm on the pristine kaolinite (Figure 2), according to the modified de Boer classification, belongs to type II (b) isotherms and are typical for nonporous sorbents with a small macroporous component [41].

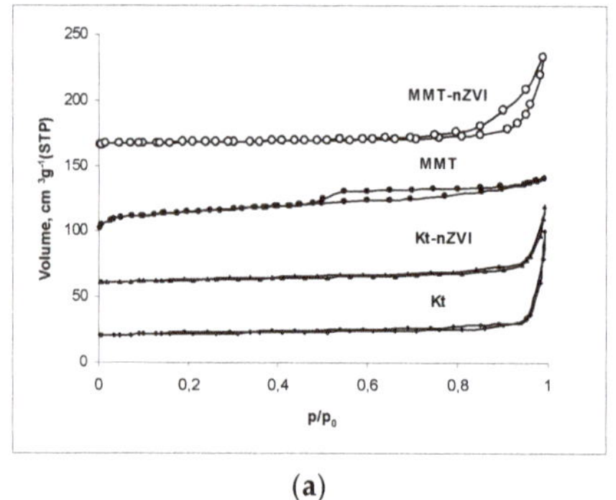

(a)

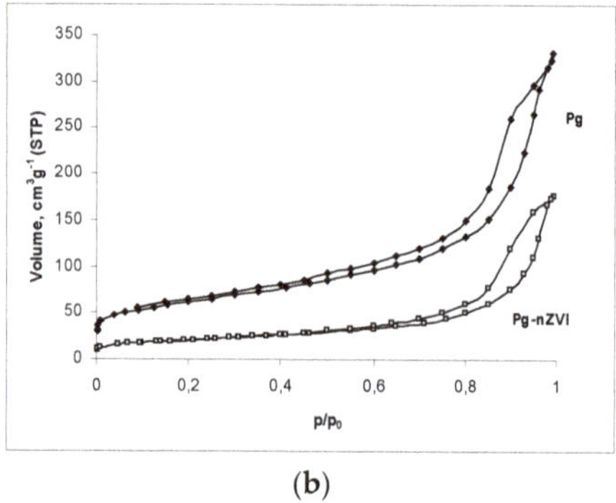

(b)

Figure 2. Nitrogen adsorption–desorption isotherms of pristine and nZVI—modified clay minerals: (**a**)—kaolinite and montmorillonite,(**b**)—palygorskite.

The narrow hysteresis loop of type H3 on isotherms is the result of capillary condensation in the structural aggregates of kaolinite between weakly interconnected flat elementary packages of the mineral. The nature of the nitrogen adsorption curves on the samples of montmorillonite and paligorskite are similar to kaolinite and, thus, also belong to type II (b) isotherms with a hysteresis loop type H3. The calculated characteristics of the porous structure of the samples are given in Table 1. As can be seen from the nitrogen isotherms, the specific surface area of samples sharply decreases after the surface modification. Such reduction is stipulated by aggregation processes of small clay particles by nZVI and practically complete closing of micropores with nZVI films and the resultant blocking of the access of nitrogen molecules to these pores.

Table 1. Characteristics of the porous structure of the samples.

Sample	S, m^2/g	V_Σ, cm^3/g	V_μ, cm^3/g	$V_{\mu\%}$, %	Distribution of Pore Size, nm			
					BJH dV (r)		DFT dV (r)	
					r_1	r_2	r_1	r_2
Kaolinite	8.98	0.124	0.003	2.1	-	-	2.36	4.91–8.58
Kaolinite-nZVI	11.72	0.093	0.003	3.1	-	-	2.51	2.63–3.52
Montmorillonite	89.11	0.081	0.016	19.8	1.41	-	1.91	2.82
Montmorillonite-nZVI	24.74	0.384	0.008	2.1	2.11	-	2.84	8.96
Palygorskite	213.15	0.512	0.084	16.6	1.90	6.26	-	-
Palygorskite-nZVI	71.68	0.274	0.002	0.8	8.89	-	6.73	8.64–9.95

Sorption (q, μmol/g) of uranium ions on pristine and modified samples are shown in Figure 3. On the kaolinite surface, sorption can occur on active centres of two types: ditrigonal siloxane wells on the basal surfaces of kaolinite particles and hydroxyl groups on broken bonds in tetrahedral (Si-O-Si) and octahedral (Al-O-Al) sheets on edge surfaces of the particles [42].

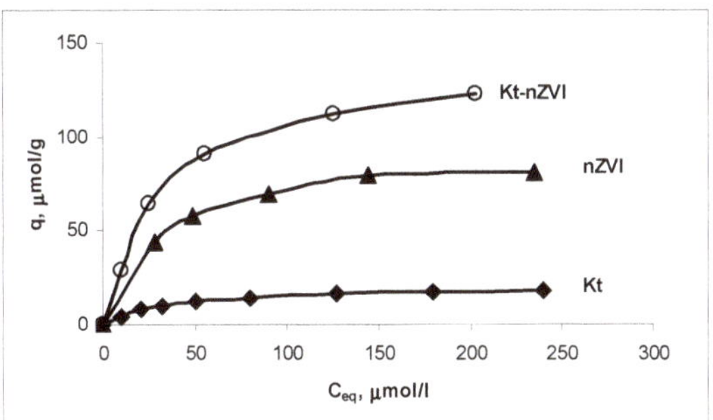

Figure 3. Sorption isotherms of U(VI) ions on pristine kaolinite and kaolinite-nZVI.

However, for the pristine Glukhovets kaolinite, due to its perfect structure and the absence of heterovalent isomorphic substitutions in the mineral structure, the charge of the structural packages is close to zero and, therefore, sorption at low concentrations of uranium ions in solution occurs mainly on hydroxyl groups located on edge surfaces of the particles.

Such sorption centres on the edge surfaces, depending on the pH of the medium, can have the composition >Si-OH and >Si-O$^-$ on the broken tetrahedral sheet of the mineral and >Al-OH$_2$$^+$; Al-OH and >Al-OH$^-$ on a broken octahedral sheet of the mineral and sorption of uranium ions on the surface occurs with the formation of strong inner-sphere surface complexes.

The results obtained for the modified samples indicate that the sorption values of uranium ions at pH 6 are several times higher than those for the pristine minerals (Figure 3). As it was established, synthesized nZVI has a core–shell structure and the thin film on the surface of the particles consists of iron oxides such as FeO, Fe_2O_3 and Fe_3O_4 and its hydroxides [14,17,43,44]. Fe(0) is in the centre of the particles. As a result of such composite structure, the removal of uranium (VI) from water by nZVI is also may be affected by the redox transformation of a highly soluble form of U(VI) to less soluble U(IV)with direct electron transfer at the Fe(0) surface [16,17,45]:

$$Fe^0 + 1.5UO_2^{2+} + 6H^+ = Fe^{3+} + 1.5\,U^{4+} + 3H_2O \quad (1)$$

In order to elucidate the synergetic effect for the Kt-nZVI composite sorbents, the removal of U(VI) by nZVI alone was also investigated. As shown in Figure 3, the removal efficiencies of U(VI) by nZVI alone wassignificantly lower than that of Kt-nZVI.

The dependence of the sorption values of uranium ions on the surface of kaolinite particles from pHis presented in Figure 4. The corresponding curves have a characteristic form with a marked minimum in the acidic and alkaline pH areas.

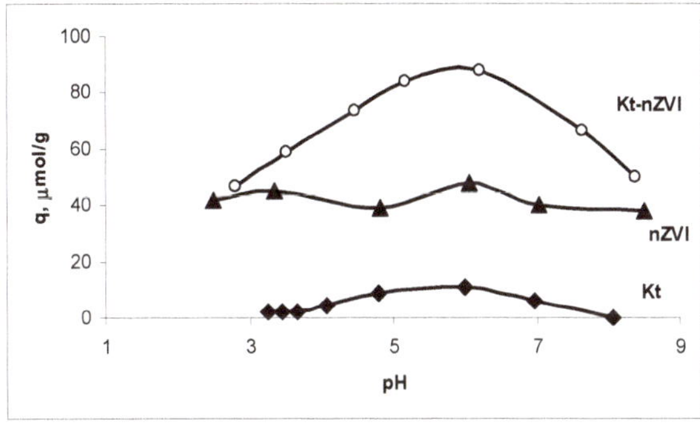

Figure 4. Sorption of U(VI) ions on pristine kaolinite and kaolinite-nZVI as a function of pH.

The last is due to the peculiarities of the chemistry of the clay surface and the complex chemistry of aqueous solutions of uranium salts [46]. In the acidic pH area, the sorption characteristics of clays in relation to uranyl cations UO_2^{2+} are determined by the neutral and positive surface charge resulting from protonation of surface groups and the formation of >>Si-OH, >Al-OH and $>Al\text{-}OH_2^+$ groups. In the alkaline area, dissociated $>Si\text{-}O^-$ and $>Al\text{-}O^-$groups predominate on the clay surface, and uranium in solution exists mainly in the form of neutral $UO_2(OH)_2$, UO_2CO_3 and negatively charged species $(UO_2)_2CO_3(OH)_3^-$ [7,47]. This causes insignificant values of sorption of uranium compounds in these pH areas, because positively charged uranium species which predominated in the acidic pH areas are not adsorbed onto the positively charged clay surface, and negatively charged uranium species in the alkaline pH areas are not adsorbed onto the negatively charged clay surface. In the neutral pH area, the charge of the kaolinite surface and the charge of uranium ions have opposite signs, which influences the proceedings of sorption processes and the appearance of maximum on the curves of the dependence of sorption values from pH.

Montmorillonite has a labile structure with the possibility of significant expansion of the inter-package distance during the penetration of polar water molecules between flat structural packages. This determines the availability for ion exchange processes not only of the outer but also of the inner surface of the particles of this clay mineral, which determines a significant increase in the sorption of uranium ions compared to kaolinite (see Figure 5). The nature of the dependence of sorption values from pH for montmorillonite corresponds to that for kaolinite, however, is less pronounced (see Figure 6). This is due to the smaller relative contribution of the edge surfaces of this mineral to its total specific surface area and the increased role of sorption centres on the basal surfaces of particles (ditrigonal wells) whose behavior is independent of pH [42].

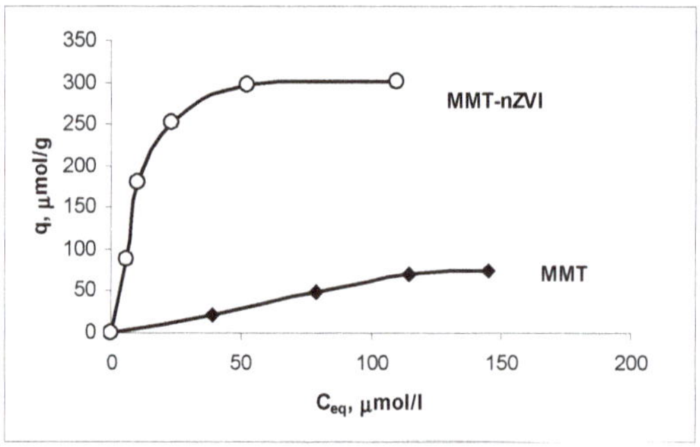

Figure 5. Sorption isotherms of U(VI) ions on pristine montmorillonite and montmorillonite-nZVI.

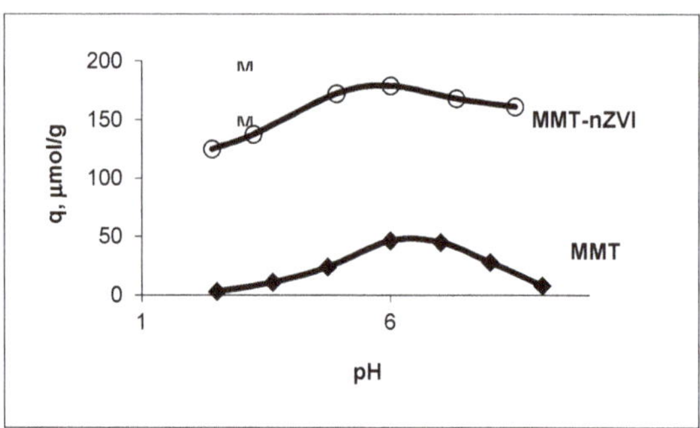

Figure 6. Sorption of U(VI) ions on pristine montmorillonite and montmorillonite-nZVI as a function of pH.

Particles of chain silicate paligorskite have an elongated rod-like shape, the volume of which is pierced by zeolite-like channels [48]. Sorption of ions occurs on the outer surface of the mineral, the value of which is greater than that of kaolinite and montmorillonite established by N_2 sorption. However, montmorillonite particles show an outstanding property to delaminate into individual silicate layers or thin packets of them in solutions. So, the values of the specific surface area of dispersions of montmorillonite samples in solutions are much higher than of air-dried samples (maximum crystallographic value of SSA are ~750–780 m^2/g [41]). Therefore, the values of the sorption of uranium on both the pristine and modified minerals are the average values between those for kaolinite and montmorillonite (see Figures 7 and 8).

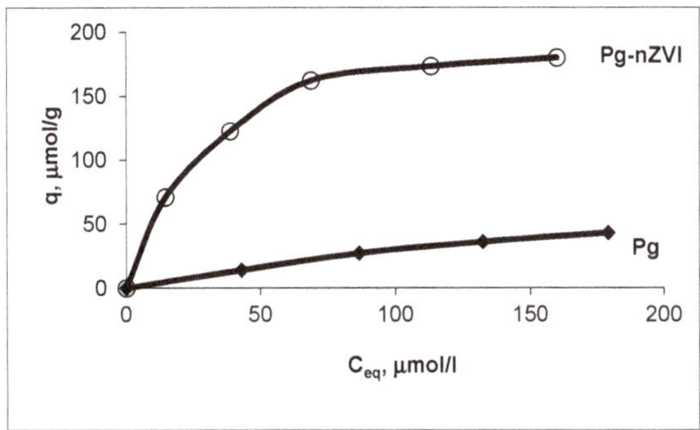

Figure 7. Sorption isotherms of U(VI) ions on pristine paligorskite and paligorskite-nZVI.

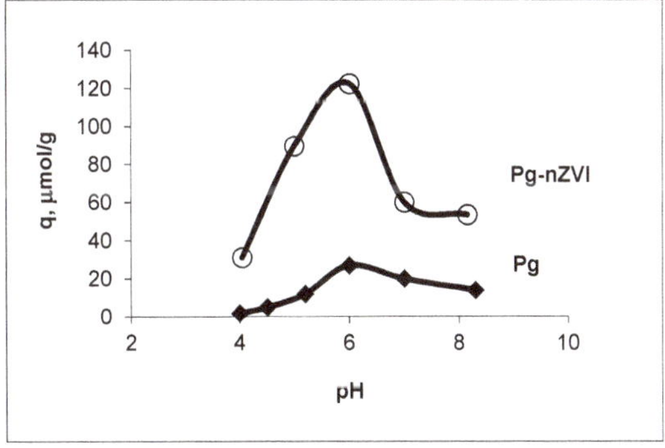

Figure 8. Sorption of U(VI) ions on pristine paligorskite and paligorskite-nZVI as a function of pH.

When considering the sorption processes in mineralized waters, it is important to analyze the forms in which uranium exists under these conditions. *Medusa* software, which is widely used in analytical practice, was used for calculations [49]. The main solid phases of uranium in aqueous systems are insoluble hydrates $UO_3 \cdot H_2O$ or $UO_2(OH)_2$ (lgK_{sp}= −20.34 – −23.5) [47,50] and carbonate UO_2CO_3 (lgK_{sp}= −13.21 – −14.26) [51].At the same time, sulfate, carbonate, phosphate and nitrate complexes of uranium may be the main species in water.The composition of main anions of underground mineralized water near the liquid waste storage facility for hydrometallurgical processing uranium ores at the Centre of Ukrainian Uranium Industry (Zhovti Vody)are: HCO_3^-—450; Cl^-—180; SO_4^{2-}—2830; NO_3^-—130 mg/L. The high affinity of uranyl ions to the nitrate is known, but the content of nitrate-ions in Zhovti Vody mineralized water is much smaller thanthe content of carbonate-ions and sulfate-ions. Therefore, we did not include nitrate-ions complexes in the speciation calculations. Zhovti Vody-mineralized waters, even with a sufficiently high content of uranium in them, are characterized by almost complete binding of uranium to sulfate complexes UO_2SO_4

and $UO_2(SO_4)_2^{2-}$ in the acidic area, as well as in carbonate complexes UO_2CO_3, $UO_2(CO_3)_2^{2-}$ and $UO_2(CO_3)_3^{4-}$ in neutral and alkaline areas (see Figure 9).

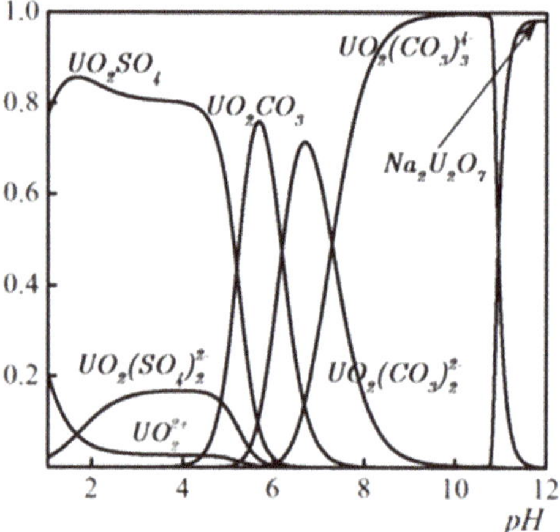

Figure 9. U(VI) speciation in mineralized groundwater.

Sorption processes in mineralized waters were studied using montmorillonite and paligorskite. The pH dependence curves of uranium sorption values have a characteristic form for both types of clays with a maximum in the pH area of 5–7 (Figure 10).

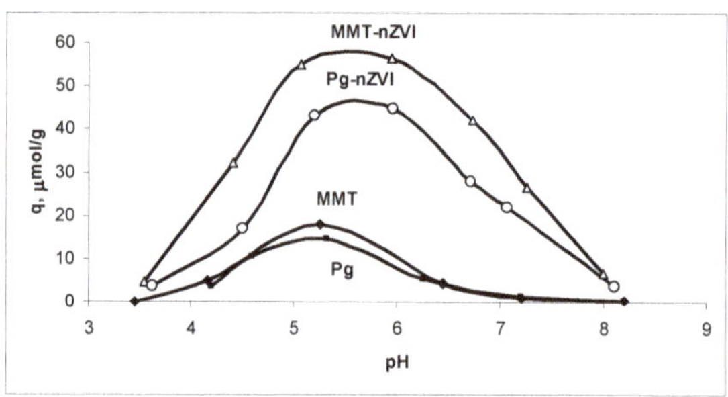

Figure 10. Sorption of U(VI) ions on pristine and nZVI-modified montmorillonite and palygorskite from mineralized waters as a function of pH.

Sorption of uranium complexes occurs primarily due to the exchange of hydroxyl ions of the hydroxide film on the surface of nanoscale iron particles:

$$M\text{-}(OH)_n + UO_2(SO_4)_2{}^{2-} \rightarrow M\text{-}(OH)_{n\text{-}2}UO_2(SO_4)_2 + 2OH^- \quad (2)$$

$$M\text{-}(OH)_n + UO_2(CO_3)_2{}^{2-} \rightarrow M\text{-}(OH)_{n\text{-}2}UO_2(CO_3)_2 + 2OH^- \quad (3)$$

When considering the mechanism of sorption of uranium compounds on the surface of clays, especially of rather high concentration of uranium, the possibility of precipitation of sparingly soluble hydroxides and others(hexavalent uranium compounds schoepite, carnotite, tyuyamunite, etc. [52]) cannot be ruled out. In mineralized waters, in contrast to diluted waters, the solubility of insoluble salts of uranium is significantly increased. Therefore, at concentrations of salts corresponding to those in mineralized waters from 3–5 to 10–20 g/L [52], the precipitation of uranium solid phases is not expected and the most probable mechanism of uranium binding is only the sorption mechanism.

Sorption isotherms of uranium from mineralized waters were obtained at pH 7.2, which corresponds to the pH value of real groundwater (Figure 11).

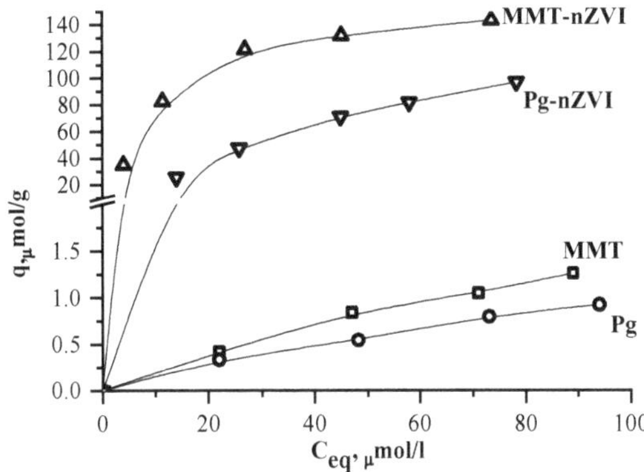

Figure 11. Sorption isotherms of U(VI) ions on pristine and nZVI-modified montmorillonite and palygorskite from mineralized waters.

Sorption isotherms were analyzed using Langmuir and Freundlich equations. The results of the calculations of the corresponding coefficients are shown in Table 2.

Table 2. Langmuir and Freundlich parameters for the adsorption of uranyl ions onto pristine and nZVI-modified minerals.

Sample	MMT	MMT-nZVI	Pg	Pg-nZVI
Langmuir				
a_∞, μmol g^{-1}	125	452	59	325
K_L, L μmol^{-1}	0.0084	0.1093	0.0096	0.1221
R^2	0.956	0.964	0.975	0.938
Freundlich				
K_F	5.3	92.0	3.2	116.5
1/n	1.97	3.28	2.13	3.35
R^2	0.887	0.825	0.850	0.892

The obtained isotherms are well described by the equation of monomolecular Langmuir sorption (correlation coefficient R^2 = 0.968–0.996) which assumes the energy homogeneity of the active centres accordingly. For the empirical Freundlich equation, which is suitable mainly for describing the starting areas of isotherms, the correlation coefficient is lower (R^2 = 0.934–0.991).

4. Conclusions

Thus, different in structure clays are effective cheap matrices for obtaining efficient sorption materials based on zero valence iron for purification of uranium-contaminated surface and mineralized groundwater, which is typical for areas of uranium ore mining and processing.In terms of sorption capacity, the composite samples form a sequence: montmorillonite-nZVI > paligorskite-nZVI > kaolinite-nZVI, which corresponds to a decrease in the specific surface area of the pristine clay minerals.

Removal of U(VI) from mineralized waters occurs primarily due to the exchange of hydroxyl ions of the hydroxide film on the surface of nanoscale iron particles by uranium sulfate and carbonate complexes.

Another possible mechanism of immobilization of uranium compounds is the reduction of uranium (VI) to uranium (IV) by electron transfer from the volume of nZVI particles through a hydroxide film to their surface with the formation and deposition of much less soluble compounds of the latter.However, it is necessary to specify the main reactions in the removal of uranium by immobilized ZVI for its further practical applications.

Author Contributions: Conceptualization, B.K.; methodology, I.K. and V.T.; validation, B.K. and I.K.; formal analysis, I.K.; investigation, I.K., V.T.; writing-original draft preparation, B.K., S.U.; writing—review & editing, B.K. All authors have read and agreed to the published version of the manuscript.

Funding: This research received no external funding.

Acknowledgments: Author is grateful to the staff of Institute for Sorption and Problems of Endoecology and Igor Sikorsky Kyiv Polytechnic Institute for the help and support.

Conflicts of Interest: The authors declare no conflict of interest.

References

1. Merkel, B.; Schipek, M. (Eds.) *The New Uranium Mining Boom*; Springer: Berlin/Heidelberg, Germany, 2011; 848p, ISBN 978-3-642-22122-4.
2. Liu, B.; Peng, T.; Hong-Juan, S.; Yue, H. Release behavior of uranium in uranium mill tailings under environmental conditions. *J. Environ. Radioact.* **2017**, *171*, 160–168. [CrossRef] [PubMed]
3. Selvakumar, R.; Ramadoss, G.; Menon, M.P.; Rajendran, K.; Thavamani, P.; Naidu, R.; Megharaj, M. Challenges and complexities in remediation of uranium contaminated soils: A review. *J. Environ. Radioact.* **2018**, *192*, 592–603. [CrossRef] [PubMed]
4. Yin, M.; Sun, J.; Chen, Y.; Wang, J.; Shang, J.; Belshaw, N.; Shen, C.; Liu, J.; Li, H.; Linghu, W.; et al. Mechanism of uranium release from uranium mill tailings under long-term exposure to simulated acid rain: Geochemical evidence and environmental implication. *Environ. Pollut.* **2019**, *244*, 174–181. [CrossRef] [PubMed]
5. Neiva, A.; Antunes, I.; Carvalho, P.C.S.; Santos, A. Uranium and arsenic contamination in the former Mondego Sul uranium mine area, Central Portugal. *J. Geochem. Explor.* **2016**, *162*, 1–15. [CrossRef]
6. Bernhard, G.; Geipel, G.; Brendler, V.; Nitsche, H. Uranium speciation in waters of different uranium mining areas. *J. Alloy. Compd.* **1998**, 201–205. [CrossRef]
7. Mühr-Ebert, E.; Wagner, F.; Walther, C. Speciation of uranium: Compilation of a thermodynamic database and its experimental evaluation using different analytical techniques. *Appl. Geochem.* **2019**, *100*, 213–222. [CrossRef]
8. Myasoyedova, G.V.; Nikashyna, V.A. Sorption materials fort he extraction of radionuclides from aqueous media. *Russ. Chem. J.* **2006**, *50*, 55–63.
9. Krajňák, A.; Viglašová, E.; Galamboš, M.; Krivosudský, L. Application of HDTMA-intercalated bentonites in water waste treatment for U(VI) removal. *J. Radioanal. Nucl. Chem.* **2017**, *314*, 2489–2499. [CrossRef]
10. Krajnák, A.; Viglašová, E.; Galamboš, M.; Krivosudský, L. Kinetics, thermodynamics and isotherm parameters of uranium(VI) adsorption on natural and HDTMA-intercalated bentonite and zeolite. *Desalin. Water Treat.* **2018**, *127*, 272–281. [CrossRef]
11. Kornilovych, B.; Wireman, M.; Caruso, B.; Koshik, Y.; Pavlenko, V.; Tobilko, V. The use of permeable reactive barrier against contaminated groundwater in Ukraine. *Central Eur. J. Occup. Environ. Med.* **2009**, *15*, 73–85.
12. Interstate Technology & Regulatory Council. *Permeable Reactive Barriers: Technology Update. PRB-5*; Interstate Technology & Regulatory Council, PRB; Technology Update Team: Washington, DC, USA, 2011; Available online: www.itrcweb.org (accessed on 17 September 2020).
13. Zhang, W.-X. Nanoscale Iron Particles for Environmental Remediation: An Overview. *J. Nanoparticle Res.* **2003**, *5*, 323–332. [CrossRef]
14. Fu, F.; Dionysiou, D.; Liu, H. The use of zero-valent iron for groundwater remediation and wastewater treatment: A review. *J. Hazard. Mater.* **2014**, *267*, 194–205. [CrossRef]
15. Zou, Y.; Wang, X.; Khan, A.; Wang, P.; Liu, Y.; Alsaedi, A.; Hayat, T.; Wang, X. Environmental Remediation and Application of Nanoscale Zero-Valent Iron and Its Composites for the Removal of Heavy Metal Ions: A Review. *Environ. Sci. Technol.* **2016**, *50*, 7290–7304. [CrossRef] [PubMed]
16. Chen, A.; Shang, C.; Shao, J.; Zhang, J.; Huang, H. The application of iron-based technologies in uranium remediation: A review. *Sci. Total. Environ.* **2017**, *575*, 1291–1306. [CrossRef]

17. Jing, C.; Li, Y.; Landsberger, S. Review of soluble uranium removal by nanoscale zero valent iron. *J. Environ. Radioact.* **2016**, *164*, 65–72. [CrossRef] [PubMed]
18. Jiang, D.; Zeng, G.; Huang, D.; Chen, M.; Zhang, C.; Huang, C.; Wan, J. Remediation of contaminated soils by enhanced nanoscale zero valent iron. *Environ. Res.* **2018**, *163*, 217–227. [CrossRef]
19. Tosco, T.; Papini, M.P.; Viggi, C.C.; Sethi, R. Nanoscale zerovalent iron particles for groundwater remediation: A review. *J. Clean. Prod.* **2014**, *77*, 10–21. [CrossRef]
20. Zhao, X.; Liu, W.; Cai, Z.; Han, B.; Qian, T.; Zhao, D. An overview of preparation and applications of stabilized zero-valent iron nanoparticles for soil and groundwater remediation. *Water Res.* **2016**, *100*, 245–266. [CrossRef] [PubMed]
21. Trujillo-Reyes, J.; Peralta-Videa, J.; Gardea-Torresdey, J.L. Supported and unsupported nanomaterials for water and soil remediation: Are they a useful solution for worldwide pollution? *J. Hazard. Mater.* **2014**, *280*, 487–503. [CrossRef]
22. Liu, Z.; Zhang, F.; Hoekman, S.K.; Liu, T.; Gai, C.; Peng, N. Homogeneously Dispersed Zerovalent Iron Nanoparticles Supported on Hydrochar-Derived Porous Carbon: Simple, in Situ Synthesis and Use for Dechlorination of PCBs. *ACS Sustain. Chem. Eng.* **2016**, *4*, 3261–3267. [CrossRef]
23. Petala, E.; Dimos, K.; Douvalis, A.; Bakas, T.; Tucek, J.; Zbořil, R.; Karakassides, M.A. Nanoscale zero-valent iron supported on mesoporous silica: Characterization and reactivity for Cr(VI) removal from aqueous solution. *J. Hazard. Mater.* **2013**, *261*, 295–306. [CrossRef] [PubMed]
24. Sun, Z.; Zheng, S.; Ayoko, G.A.; Frost, R.L.; Xi, Y. Degradation of simazine from aqueous solutions by diatomite-supported nanosized zero-valent iron composite materials. *J. Hazard. Mater.* **2013**, *263*, 768–777. [CrossRef]
25. Yua, S.; Wangab, X.; Liua, Y.; Chena, Z.; Wua, Y.; Liua, Y.; Panga, H.; Songc, G.; Chen, J.; Wang, X. Efficient removal of uranium(VI) by layered double hydroxides supported nanoscale zero-valent iron: A combined experimental and spectroscopic studies. *Chem. Eng. J.* **2019**, *365*, 51–59. [CrossRef]
26. Ding, C.; Cheng, W.; Nie, X.; Yi, F.; Xiang, S.; Asiri, A.M.; Marwani, H.M. Reactivity of carbonized fungi supported nanoscale zero-valent iron toward U(VI) influenced by naturally occurring ions. *J. Ind. Eng. Chem.* **2018**, *61*, 236–243. [CrossRef]
27. Lv, X.; Xue, X.; Jiang, G.; Wu, D.; Sheng, T.; Zhou, H.; Xinhua, X. Nanoscale Zero-Valent Iron (nZVI) assembled on magnetic Fe3O4/graphene for Chromium (VI) removal from aqueous solution. *J. Colloid Interface Sci.* **2014**, *417*, 51–59. [CrossRef] [PubMed]
28. Tan, L.; Wang, Y.; Liu, Q.; Wang, J.; Jing, X.; Liu, L.; Liu, J.; Song, D. Enhanced adsorption of uranium (VI) using a three-dimensional layered double hydroxide/graphene hybrid material. *Chem. Eng. J.* **2015**, *259*, 752–760. [CrossRef]
29. Kurniawan, T.A.; Chan, G.Y.; Lo, W.-H.; Babel, S. Comparisons of low-cost adsorbents for treating wastewaters laden with heavy metals. *Sci. Total. Environ.* **2006**, *366*, 409–426. [CrossRef]
30. Misaelides, P. Clay minerals and zeolites for radioactive waste immobilization and containment. *Modif. Clay Zeolite Nanocomposite Mater.* **2019**, 243–274. [CrossRef]
31. Ezzatahmadi, N.; Ayoko, G.A.; Millar, G.J.; Speight, R.; Yan, C.; Li, J.; Li, S.; Zhu, J.; Xi, Y. Clay-supported nanoscale zero-valent iron composite materials for the remediation of contaminated aqueous solutions: A review. *Chem. Eng. J.* **2017**, *312*, 336–350. [CrossRef]
32. Üzüm, Ç.; Shahwan, T.; Eroğlu, A.E.; Hallam, K.R.; Scott, T.B.; Lieberwirth, I. Synthesis and characterization of kaolinite-supported zero-valent iron nanoparticles and their application for the removal of aqueous Cu^{2+} and Co^{2+} ions. *Appl. Clay Sci.* **2009**, *43*, 172–181. [CrossRef]
33. Shahwan, T.; Üzüm, Ç.; Eroğlu, A.E.; Lieberwirth, I. Synthesis and characterization of bentonite/iron nanoparticles and their application as adsorbent of cobalt ions. *Appl. Clay Sci.* **2010**, *47*, 257–262. [CrossRef]
34. Bhowmick, S.; Chakrabarti, S.; Mondal, P.; Van Renterghem, W.; Berghe, S.V.D.; Roman-Ross, G.; Chatterjee, D.; Iglesias, M. Montmorillonite-supported nanoscale zero-valent iron for removal of arsenic from aqueous solution: Kinetics and mechanism. *Chem. Eng. J.* **2014**, *243*, 14–23. [CrossRef]
35. Jing, C.; Landsberger, S.; Li, Y. The application of illite supported nanoscale zero valent iron for the treatment of uranium contaminated groundwater. *J. Environ. Radioact.* **2017**, *175*, 1–6. [CrossRef]
36. Zhdanyuk, N.V.; Kovalchuk, I.A.; Kornilovych, B.Y. Sorption of uranium (VI) and cobalt (II) ions by iron-containing nanocomposites based on palygorskite. *Surface Chem. Phys. Techn.* **2019**, *10*, 48–58. [CrossRef]

37. Rouquerol, F.; Rouquerol, J.; Sing, K.S.W.; Llewellyn, P.; Maurin, G. *Adsorption by Powders and Porous Solids*; Elsevier: Amsterdam, The Netherlands, 2014; ISBN 9780080970356.
38. Shi, L.-N.; Lin, Y.-M.; Zhang, X.; Chen, Z.-L. Synthesis, characterization and kinetics of bentonite supported nZVI for the removal of Cr(VI) from aqueous solution. *Chem. Eng. J.* **2011**, *171*, 612–617. [CrossRef]
39. Tobilko, V.Y.; Kornilovych, B.Y. Synthethis and sorption properties of composite materials based on nanoscale Fe. *East. Eur. J. Enterp. Technol.* **2015**, *4*, 22–27.
40. Kornilovych, B.; Wireman, M.; Ubaldini, S.; Guglietta, D.; Koshik, Y.; Caruso, B.; Kovalchuk, I. Uranium Removal from Groundwater by Permeable Reactive Barrier with Zero-Valent Iron and Organic Carbon Mixtures: Laboratory and Field Studies. *Metals* **2018**, *8*, 408. [CrossRef]
41. Bergaya, F.; Theng, B.K.; Lagaly, G. *Handbook of Clay Science*; Elsevier: London, UK, 2006; 1246p, ISBN 9780080441832.
42. Sposito, G. *The Surface Chemistry of Soils*; Oxford University Press: Oxford, UK, 1984; 344p, ISBN 9780195313697.
43. Crane, R.; Pullin, H.; Scott, T. The influence of calcium, sodium and bicarbonate on the uptake of uranium onto nanoscale zero-valent iron particles. *Chem. Eng. J.* **2015**, *277*, 252–259. [CrossRef]
44. Chekli, L.; Bayatsarmadi, B.; Sekine, R.; Sarkar, B.; Shen, A.M.; Scheckel, K.G.; Skinner, W.; Naidu, R.; Shon, H.K.; Lombi, E.; et al. Analytical characterisation of nanoscale zero-valent iron: A methodological review. *Anal. Chim. Acta* **2016**, *903*, 13–35. [CrossRef]
45. Sheng, G.; Shao, X.; Li, Y.; Li, J.; Dong, H.; Cheng, W.; Gao, X.; Huang, Y. Enhanced Removal of Uranium(VI) by Nanoscale Zerovalent Iron Supported on Na–Bentonite and an Investigation of Mechanism. *J. Phys. Chem. A* **2014**, *118*, 2952–2958. [CrossRef]
46. Kornilovich, B.Y.; Pshinko, G.N.; Bogolepov, A.A. Effects of EDTA and NTA on sorption of U(VI) on the clay fraction of soil. *Radiochemistry* **2006**, *48*, 584–588. [CrossRef]
47. Altmaier, M.; Yalçıntaş, E.; Gaona, X.; Neck, V.; Müller, R.; Schlieker, M.; Fanghänel, T. Solubility of U(VI) in chloride solutions. I. The stable oxides/hydroxides in NaCl systems, solubility products, hydrolysis constants and SIT coefficients. *J. Chem. Thermodyn.* **2017**, *114*, 2–13. [CrossRef]
48. Galan, E.; Singer, A. (Eds.) *Developments in Palygorskite-Sepiolite Research*; Galan, E.; Singer, A. (Eds.) Elsevier: Amsterdam, The Netherlands, 2011; Volume 3, 520p, ISBN 9780444536075.
49. Puigdomènech, I.; Colas, E.; Glive, M.; Campos, I.; Garcia, D. A tool to draw chemical equilibrium diagrams using SIT: Applications to geochemical systems and radionuclide solubility. *MRS Online Proc. Libr. Arch.* **2014**, *1665*, 111–116. [CrossRef]
50. Fujiwara, K.; Yamana, H.; Fujii, T.; Kawamoto, K.; Sasaki, T.; Moriyama, H. Solubility Product of Hexavalent Uranium Hydrous Oxide. *J. Nucl. Sci. Technol.* **2005**, *42*, 289–294. [CrossRef]
51. Kramer-Schnabel, U.; Bischoff, H.; Xi, R.H.; Marx, G. Solubility Products and Complex Formation Equilibria in the Systems Uranyl Hydroxide and Uranyl Carbonate at 25°C and I = 0.1 M. *Radiochim. Acta* **1992**, *56*, 183–188. [CrossRef]
52. Prentice Hall; Langmuir, D. *Aqueous Environmental Geochemistry*; Prentice Hall: Upper Saddle River, NJ, USA, 1997; 600p, ISBN 13-978-0023674129.

Publisher's Note: MDPI stays neutral with regard to jurisdictional claims in published maps and institutional affiliations.

Article

Leaching Kinetics of Sulfides from Refractory Gold Concentrates by Nitric Acid

Denis A. Rogozhnikov, Andrei A. Shoppert *, Oleg A. Dizer, Kirill A. Karimov and Rostislav E. Rusalev

Department of Non-ferrous Metals Metallurgy, Ural Federal University, Yekaterinburg 620002, Russia; darogozhnikov@yandex.ru (D.A.R.); oleg.dizer@yandex.ru (O.A.D.); kirill_karimov07@mail.ru (K.A.K.); rusalevrostislav@gmail.com (R.E.R.)

* Correspondence: andreyshop@list.ru; Tel.: +7-922-024-3963

Received: 5 April 2019; Accepted: 19 April 2019; Published: 22 April 2019

Abstract: The processing of refractory gold-containing concentrates by hydrometallurgical methods is becoming increasingly important due to the depletion of rich and easily extracted mineral resources, as well as due to the need to reduce harmful emissions from metallurgy, especially given the high content of arsenic in the ores. This paper describes the investigation of the kinetics of HNO_3 leaching of sulfide gold-containing concentrates of the Yenisei ridge (Yakutia, Russia). The effect of temperature (70–85 °C), the initial concentration of HNO_3 (10–40%) and the content of sulfur in the concentrate (8.22–22.44%) on the iron recovery into the solution was studied. It has been shown that increasing the content of S in the concentrate from 8.22 to 22.44% leads to an average of 45% increase in the iron recovery across the entire range temperatures and concentrations of HNO_3 per one hour of leaching. The leaching kinetics of the studied types of concentrates correlates well with the new shrinking core model, which indicates that the reaction is regulated by interfacial diffusion and diffusion through the product layer. Elemental S is found on the surface of the solid leach residue, as confirmed by XRD and SEM/EDS analysis. The apparent activation energy is 60.276 kJ/mol. The semi-empirical expression describing the reaction rate under the studied conditions can be written as follows: $1/3\ln(1 - X) + [(1 - X)^{-1/3} - 1] = 87.811(HNO_3)^{0.837}(S)^{2.948}e^{-60276/RT}\cdot t$.

Keywords: refractory gold concentrate; resources depletion; reducing harmful emissions; arsenic; nitric acid; kinetics; shrinking core model; pyrite; arsenopyrite

---

1. Introduction

Russia possesses large reserves of gold—more than 14 thousand tons—which exceeds the reserves of the world's main producers, China and Australia, and is slightly inferior to South Africa and Canada. The Russian Federation accounts for 8% of the total world gold production and is among the three largest global producers of the precious metal. Gold-sulfide-quartz and gold-arsenic-sulfide deposits occupy a leading position in the structure of Russia's reserves; the quality of ores is comparable to the world objects of this type [1]. Very important are gold-polysulfide deposits, characterized by relatively high concentrations of gold (3.5–7 gpt).

At the same time, there is a global problem in the metallurgical industry that the quality of processed raw materials is deteriorating due to the depletion of mineral reserves and the extraction of the richest and most easily extractable ore layers. As a result, there is a need to engage poorer and more refractory ores, which are often not amenable to traditional enrichment methods.

The deterioration of ore quality, especially with the transition to the production of lower horizons, occurs in terms of reduction of metal content as well as in terms of increasing proportion of ores with fine and emulsive impregnation of sulfides in one another and the latter in waste minerals.

The share of gold ores of non-ferrous metals, where gold is an associated valuable component, accounts for 18% of the global reserves [2]. Among them, a special place belongs to ores from which gold cannot be extracted by traditional technologies.

Refractory gold-containing ores refer to materials, the extraction of gold from which by cyanidation is low or requires significant amounts of energy and reagents [3].

Currently, it is considered proven that the refractory characteristics of gold associated with sulfides is not only due to the presence of nanoparticles of native gold [4–7], but also due to the existence of solid solution, colloidal particles, surface gold [8–11]. The size of such "invisible" gold may be on the order of nanometers, which explains why it is impossible to extract it by cyanidation, even with the use of ultrafine grinding. It has also been established that in pyrite, which is one of the main carriers of gold in refractory ores, the content of "invisible" gold is greater proportionately to the higher content of arsenic in pyrite and to the finer grain. For example, in the ores of the Twin Creeks deposit, relatively coarse-grained pyrite (10–30 μm) is associated with the lowest arsenic and gold contents (less than 1% As and 17–60 g/t Au), while fine-grained pyrites (less than 2 μm) are the highest (1–2.4% As and 600–1500 g/t Au) [12].

Also important in the detection of "invisible" gold in pyrite grains is the uneven distribution of As and Au over the grain section. A thin peripheral layer of pyrite grain is enriched with arsenic, forming the so-called arsenic pyrite $Fe(As,S)_2$. It tends to contain most of the gold [13].

The nature of the chemical bond of gold, which is in the form of a solid solution in arsenic pyrite, has not been fully established and is the subject of discussion [12,14].

Invisible gold may also exist in arsenopyrite in the simple form of nanoparticles (Au^0) and in the oxidized state (Au^{1+}), and their ratio may vary significantly. For example, in the arsenopyrite deposits Jinya (China) [15], Elmtree (Canada) [16], Sao Bento (Brazil) [8] and Sheba (South Africa) [8,16] solid gold (Au^{1+}) is the predominant form as compared with nanogold (Au^0). On the contrary, arsenopyrite of the Olimpiadinskoe deposit (Russia) [17] contains "invisible" gold mainly in the form of nanoparticles (Au^0) [10].

Traditional methods of processing such refractory materials consist in the oxidation of gold-containing minerals (pyrite or arsenopyrite) in order to destroy their crystal lattice and release the gold particles by oxidative roasting [18]. This process is associated with the oxidation of iron-containing sulfides and converting arsenic to the gaseous phase. At the same time, arsenic is one of the most dangerous and carcinogenic elements [19–24] and its content in drinking water in several countries already exceeds the concentration recommended by the World Health Organization (WHO) and the United States Environmental Protection Agency (USEPA) [25,26]. Therefore, hydrometallurgical methods of sulfide oxidation have been widely implemented in recent years. The most common of them are pressure and bacterial oxidation, and leaching after fine grinding [27].

One of the possible methods of hydrometallurgical processing of refractory sulfide raw materials is the use of HNO_3 to oxidize the materials without the use of high pressure or fine grinding [28–31], as nitric acid is one of the most effective oxidizing and leaching agents [32,33].

Among the most famous technologies based on the use of HNO_3: NSC-process (nitrogen species-catalyzed pressure leaching), implemented in 1984–1995 at the Sunshine plant in the USA [34]; NITROX (in atmosphere air) and ARSENO PROCESS® (use of compressed oxygen) [35]; a subspecies of ARSENO, the REDOX PROCESS® (at above 180 °C to eliminate the formation of elemental S) [36]; the HMC process (a mixture of salts of nitric and hydrochloric acids) [37]; the Caschman process and its modification Artek Caschman, which aims to process gold-containing arsenic materials using chlorine-containing reagents [3]. However, none of these is currently used commercially for one reason or another.

Therefore, it seems relevant to conduct further studies of alternative energy saving and environmentally efficient hydrometallurgical technologies for processing sulfide gold-containing raw materials using HNO_3. There are only a few published works devoted to the theoretical aspects of HNO_3 leaching of sulfide gold-containing concentrates [38], which demonstrates the need for additional investigation using other types of concentrates with different mineralogical compositions. At the same time, there is a sufficient amount of work showing that various types of leaching reactions of raw materials with HNO_3 can be described quite accurately using the shrinking core model [39–43], which makes it possible to obtain more data on the limiting stages of the reactions.

Considering the above factors, this paper studies the kinetics of HNO_3 leaching of refractory gold-containing concentrates with the use of shrinking core models, focusing on the role of temperature, concentration of HNO_3, which, in our opinion, has not been sufficiently studied theoretically or practically. Particular attention is paid to the initial content of sulfur in the raw materials as one of the most important factors affecting the intensity, completeness and kinetic characteristics of the leaching process.

2. Materials and Methods

2.1. Analisys

The chemical analysis of the original materials and the resulting solid products of the studied processes was performed using an Axios MAX X-ray fluorescence spectrometer (XRF) (Malvern Panalytical Ltd., Almelo, Netherlands). The chemical analysis of the obtained solutions was performed by mass spectrometry with inductively coupled plasma (ICP-MS) using an Elan 9000 instrument (PerkinElmer Inc., Waltham, MA, USA). The phase analysis was performed on an XRD 7000 Maxima diffractometer (Shimadzu Corp., Tokyo, Japan).

Scanning electron microscopy with microprobe energy dispersive analysis was performed using a JSM-6390LV microscope (JEOL Ltd., Tokyo, Japan) with the INCA Energy 450 X-Max 80 adaptor, with an accelerating voltage of 20 kV.

The study on the distribution of gold in sulfide minerals was carried out using inductively coupled plasma mass spectrometry (LA-ICP-MS—NexION 300S quadrupole mass spectrometer, PerkinElmer Inc., Waltham, MA, USA) with laser ablation of the sample with the NWR 213 adaptor for a Jeol JSM-6390LV.

2.2. Experiments

Laboratory experiments on HNO_3 leaching were carried out on an apparatus consisting of a 2 dm^3 borosilicate glass round bottom reactor (Lenz Laborglas GmbH & Co. KG, Wertheim, Germany), with openings for injecting HNO_3, as well as for temperature control and removal of nitrous gases through a water-cooled reflux condenser. The reactor was thermostated. The materials were stirred using an overhead mixer at 400 rpm, which ensured uniform density of the pulp. A portion of the concentrate weighing 60 g was added to a prepared solution of HNO_3 of a required concentration. At the end of the experiment, the leaching pulp was filtered in a Buchner funnel (ECROSKHIM Co., Ltd., St. Petersburg, Russia); the solutions were sent for ICP-MS analysis; the leaching cake was washed with distilled water, dried at 100 °C to constant weight, weighed and sent for XRF analysis. All the experiments were performed three times and the mean values are presented here.

To trap the nitrous gases formed during the leaching process and to regenerate HNO_3, we used a system consisting of three successively connected absorption columns filled with distilled water and a sanitary column with a solution of thiourea to recover the residual amount of oxides to elemental nitrogen.

2.3. Materials and Reagents

The materials used in the study are three samples of refractory gold-containing sulfide flotation concentrates (size 90% less than 74 microns) from a Russian deposit of Yenisei ridge (Yakutia), obtained under different enrichment conditions. The compositions of the samples are presented in Table 1. The chemical agents were of analytical grade; the water had been purified by distillation using a GFL-manufactured device (GFL mbH, Burgwedel, Germany).

Table 1. Chemical composition of samples of refractory gold-containing sulfide concentrates, wt.%.

Element	Al	As	Fe	S	Si	Sb	K	Ca	Mg	Au (gpt)	Other
Concentrate-1	2.38	3.00	7.26	8.22	19.50	2.96	1.15	7.25	2.17	15–20	46.11
Concentrate-2	5.15	4.68	13.14	13.08	22.02	2.25	2.49	1.35	1.02	30	34.82
Concentrate-3	2.49	13.20	23.60	22.44	7.72	2.51	1.84	0.73	0.55	60–80	24.92

Figure 1 shows the X-ray diffraction pattern of the phase composition of the original concentrate-3. Table 2 presents the results of the mineralogical composition study of the concentrate samples. The results were obtained in X-ray phase analysis and X-ray microanalysis.

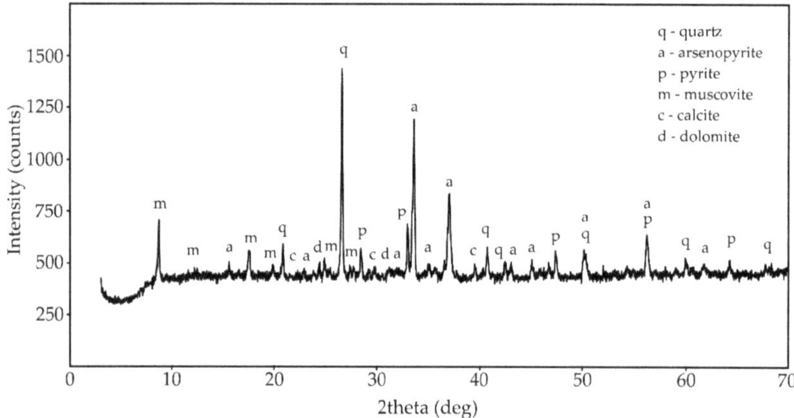

Figure 1. X-ray diffraction (XRD) pattern of the phase composition of concentrate-3.

Table 2. The results of the mineralogical composition study of the concentrate samples.

Mineral	Formula	Concentrate-1	Concentrate-2	Concentrate-3
			wt.%	
Quartz	SiO_2	36.43	35.63	10.97
Pyrite	FeS_2	10.79	20.72	29.57
Arsenopyrite	FeAsS	6.52	10.18	28.68
Stibnite	Sb_2S_3	4.13	3.15	3.50
Muscovite	$KAl_2[AlSi_3O_{10}](OH)_2$	11.70	19.15	12.24
Calcite	$CaCO_3$	9.14	-	-
Dolomite	$CaMg(CO_3)_2$	16.49	6.21	3.01
Other	-	4.80	4.96	12.03

Figure 2 presents a general view of concentrate-3 in backscattered electrons. The red dots indicate the place of LA-ICP-MS analysis; the analytical point diameter is 25 microns.

The study showed the almost complete absence of gold in antimonite. The concentration of gold in pyrite is fairly evenly distributed and does not exceed 16 gpt. The arsenopyrite distribution is uneven. The gold content ranges from 1 to 172 gpt (Table 3).

Table 3. Gold content in minerals [1] of flotation concentrate-3, gpt.

Element	Pyrite	Pyrite	Arsenopyrite	Antimonite	Arsenopyrite
Au	15.77	11.23	13.61	3.47	65.71
Element	**Arsenopyrite**	**Arsenopyrite**	**Arsenopyrite**	**Arsenopyrite**	**Arsenopyrite**
Au	1.71	172.45	26.36	12.48	0.558

[1] The materials were determined according to the content of Fe, As, Sb, S. Sulfur was used as an internal standard.

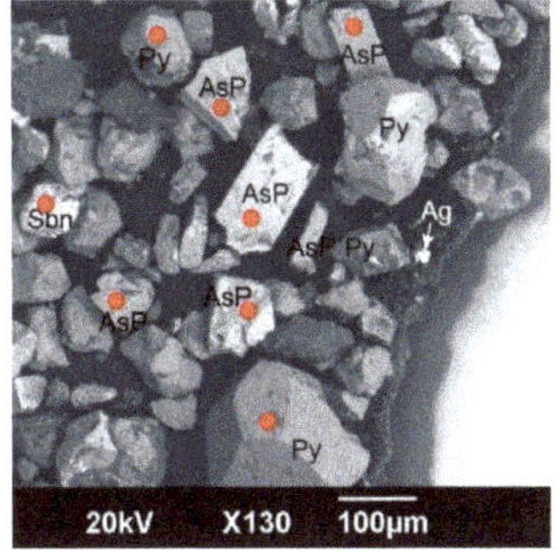

Figure 2. General view of concentrate-3 in backscattered electrons: AsP—arsenopyrite, Py—pyrite, Sbn—stibnite (antimonite), Ag—silver, dark gray grains—silicate minerals.

3. Results and Discussion

3.1. Chemistry of HNO_3 Leaching

As shown by the results of the LA-ICP-MS method with the NWR 213 adaptor, a characteristic feature of the material is that most of the gold is enclosed in finely disseminated form in the crystalline lattice of pyrite and arsenopyrite. Therefore, the main goal of the process is to reveal these two sulfide minerals. Their interaction with HNO_3 can follow these typical reactions (Equations (1)–(7)):

$$FeS_2 + 8HNO_3 = Fe(NO_3)_3 + 2H_2SO_4 + 2H_2O + 5NO, \quad (1)$$

$$2FeS_2 + 10HNO_3 = Fe_2(SO_4)_3 + H_2SO_4 + 4H_2O + 10NO, \quad (2)$$

$$FeS_2 + 18HNO_3 = Fe(NO_3)_3 + 7H_2O + 2H_2SO_4 + 15NO_2, \quad (3)$$

$$2FeS_2 + 8HNO_3 = Fe_2(SO_4)_3 + S^0 + 8NO + 4H_2O, \quad (4)$$

$$FeAsS + 17HNO_3 = Fe(NO_3)_3 + H_3AsO_4 + H_2SO_4 + 14NO_2 + 6H_2O, \quad (5)$$

$$FeAsS + 14HNO_3 = FeAsO_4 + H_2SO_4 + 14NO_2 + 6H_2O, \quad (6)$$

$$3FeAsS + 12HNO_3 = 3FeAsO_4 + 2H_2SO_4 + S^0 + 12NO + 4H_2O. \quad (7)$$

In addition, at the initial moment, NO^+ ions could be formed, which, according to Anderson et al. [44], acts as the strongest oxidizer in such systems.

Previously, we had studied the thermodynamic characteristics for the above interactions [33]. The study found that the oxidation potential of the system is to be maintained at 0.6 V or higher for reactions with transfer of iron and arsenic into the solution. A high oxidation potential is also necessary for the sulfide S^{2-} to be oxidized into sulfate $SO_4{}^{2-}$ with minimized formation of elemental S^0 that impedes a more complete oxidation of sulfide minerals and reduces subsequent recovery of gold. The results of the performed thermodynamic studies and laboratory experiments allowed us to establish the ranges of the leaching process parameters: temperature of 70–100 °C, acid concentration of 10–60% at L:S 20:1. Lower values of L:S are also possible; however, it was previously found that at low values of L:S and a high acid concentration, the reaction is very intense, which impedes temperature control, while the degree of extraction in this case changes slightly [33].

3.2. The Effect of Process Parameters on the Oxidation of Sulfides

According to Tables 1 and 2 and Figure 1, it is obvious that the concentrates studied in this work differ greatly in the content of sulfides; therefore, the study considered only the kinetics of HNO_3 leaching of each of the concentrates. The effect of temperature, HNO_3 concentration and sulfur content on iron extraction in HNO_3 leaching is shown in Figure 3. Iron recovery was considered as the main component indicating the degree of sulfide oxidation, as it is included in both pyrite and arsenopyrite. For each of the concentrates, experiments were carried out at three different HNO_3 concentrations and at four temperatures. Figure 3a–c, for example, shows the kinetic leaching curves of concentrate-1 at a concentration of nitric acid of 10, 20 and 40%, respectively, that could be used to evaluate the influence of HNO_3 concentration on leaching efficiency. It is possible to evaluate the influence of sulfur content in concentrate on the oxidation of sulfides, for example, by comparing Figure 3a,d,g.

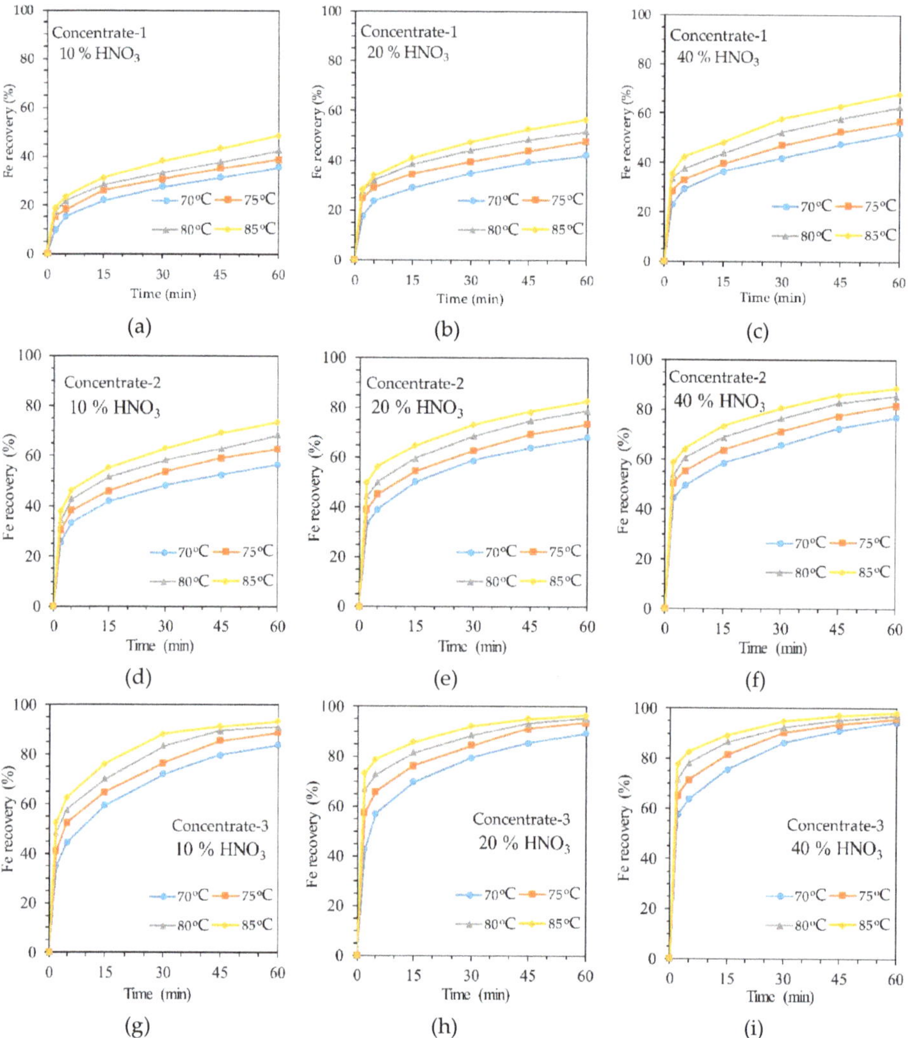

Figure 3. The dependence of iron extraction in the leaching of refractory gold-containing concentrate on temperature: (**a**) concentrate-1 and 10% HNO_3 concentration; (**b**) concentrate-1 and 20% HNO_3 concentration; (**c**) concentrate-1 and 40% HNO_3 concentration; (**d**) concentrate-2 and 10% HNO_3 concentration; (**e**) concentrate-2 and 20% HNO_3 concentration; (**f**) concentrate-2 and 40% HNO_3 concentration; (**g**) concentrate-3 and 10% HNO_3 concentration; (**h**) concentrate-3 and 20% HNO_3 concentration; (**i**) concentrate-3 and 40% HNO_3 concentration.

It can be concluded that temperature has a significant effect on iron recovery. Increasing the temperature from 70 to 85 °C effected an increase in iron extraction from concentrate-1 from 35.53 to 48.52% after 1 h of leaching with a 10% solution of HNO_3. A similar effect of temperature on iron extraction was observed for all types of concentrates and HNO_3 concentrations.

It is obvious that the increase in the concentration of nitric acid should accelerate the reaction rate of sulfide oxidation, since the excess acid facilitates the diffusion of the reagent to the reaction interface. The data on Figure 3 allows one to conclude that the concentration of HNO_3 has approximately the same effect on iron extraction as temperature. Increasing the concentration of HNO_3 from 10 to 40% for all temperatures and concentrates can increase the degree of iron extraction by an average of 20%.

The sulfur content in the concentrate has the greatest effect on the extraction of iron. Thus, an increase in sulfur content from 8 to 22% makes it possible to increase iron recovery into the solution by an average of 45% at all temperatures and concentrations of HNO_3. This is most likely due to the fact that the increase in the content of sulfides in the raw material make easier the diffusion of the reagent to them even on the last stage of the process.

Thus, particular parameters must be in place to achieve a more complete extraction of iron, and accordingly, the oxidation of sulfides for different samples of the concentrate. For example, a low-sulfur concentrate requires the maximum temperature at a high concentration of HNO_3, while for a high-sulfur concentrate, a temperature of 70 °C and a 10% HNO_3 solution is sufficient.

3.3. *Characteristics of Solid Residue*

The cake obtained as a result of HNO_3 leaching of the samples was subjected to scanning electron microscopy to study changes in the morphology of the sample in the course of the leaching process. Figure 4 shows micrographs of the original sample and the cake obtained by leaching.

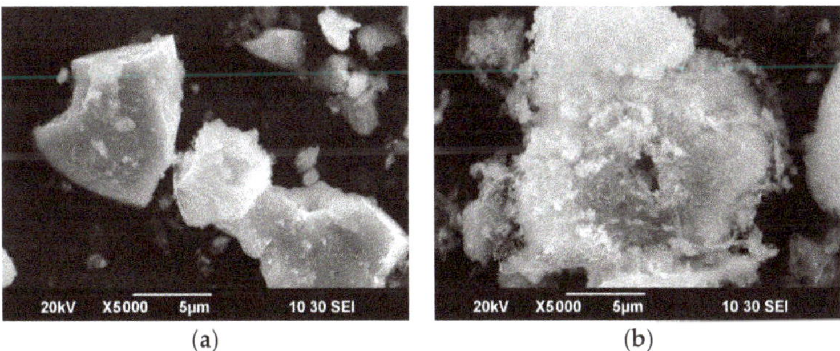

Figure 4. Scanning electron microscopy (SEM) images of the original sample of concentrate-3 (**a**) and the cake obtained by leaching (**b**).

As can be seen from Figures 2 and 4a, the original concentrate consists of particles ranging in size from 1 to 100 μm, and the surface of the particles is rather smooth. In the course of leaching (Figure 4b), a large number of cavities form on the surface of the particles, which is associated with the dissolution of sulfide minerals. Figure 5 shows micrographs of solid residue of concentrate-3 (points 001-010) and concentrate-1 (points 011-019), differing by the degree of extraction.

Energy dispersive spectroscopy was used to determine the chemical composition at different points of the samples (Table 4).

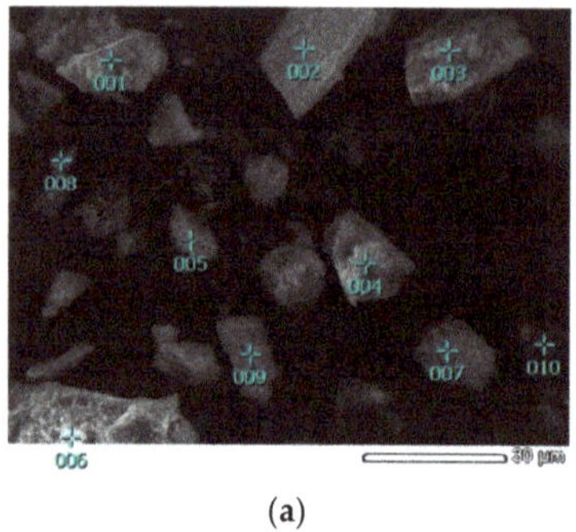

(a)

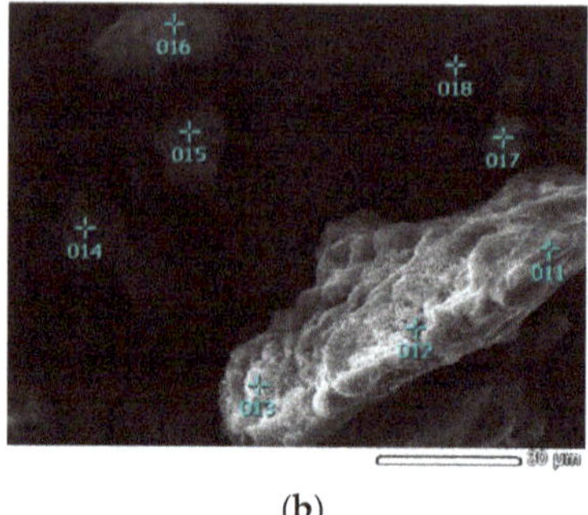

(b)

Figure 5. SEM images of solid residue with the EDS points: (**a**) solid residue of concentrate-3 leaching and (**b**) solid residue of concentrate-1 leaching.

Table 4. Results of energy dispersive spectroscopy analysis, wt.%.

Element	Al	Si	S	Fe	As	Total
Point 001	5.9	93.1	0.0	0.3	0.7	100.0
Point 003	2.6	94.4	1.0	1.3	0.6	100.0
Point 005	2.5	95.9	0.7	0.3	0.6	100.0
Point 007	7.5	85.0	1.0	3.0	3.6	100.0
Point 009	29.4	69.4	0.2	0.8	0.3	100.0
Point 0011	4.5	85.5	3.3	2.9	3.9	100.0
Point 0013	9.1	41.1	9.7	13.6	26.4	100.0
Point 0015	10.6	21.0	9.0	28.5	30.9	100.0
Point 0017	6.7	52.9	8.2	6.3	25.9	100.0

The measurement results in Table 4 confirm the presence of a large amount of unoxidized sulfides in the cake of the first sample and the almost complete absence of sulfur in the second cake, which agrees well with the results of the analysis of the liquid phase. Therefore, the surface of the particles of the concentrate-3 cake does not have a layer formed by the reaction product, elemental sulfur, which is also confirmed by the results of X-ray phase analysis, as shown in Figure 6. The absence of the reaction product on the surface of the concentrate-3 solid residue also explains the faster kinetics of leaching of the concentrate with high sulfur content because diffusion limitation is lower in this case. The absence of elemental sulfur could be attributed to the high intensity of the process, amount of the heat of exothermic reactions and the amount of gas produced that can lead to a higher oxidation potential.

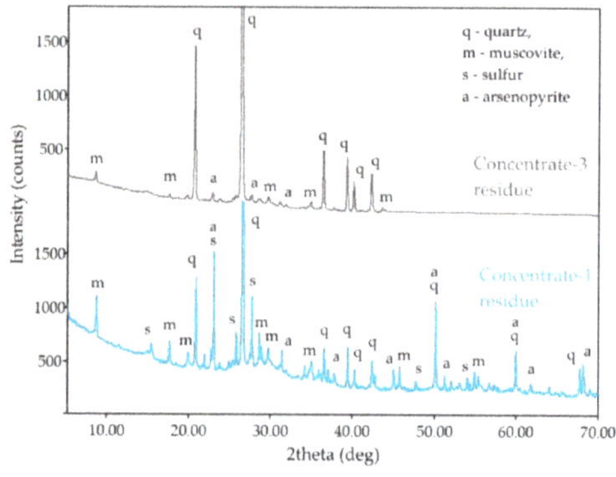

Figure 6. XRD patterns of the solid residue of the concentrate-3 and concentrate-1.

3.4. Kinetic Model

Since the degree of sulfide oxidation is greatly influenced by the temperature and concentration of HNO_3, the leaching of refractory gold-containing concentrates with HNO_3 can be controlled by diffusion as well as by kinetic stages. That is, the slowest stage can be the reagent diffusion towards the reaction surface as well as the chemical reaction itself. To determine the limiting step, it is necessary to conduct a study of the kinetics.

The shrinking core model (SCM) is generally used to describe the kinetics of heterogeneous reactions involving non-porous materials. The SCM assumes that the process rate is controlled either by the diffusion of the reactant to the surface through the diffusion layer (a liquid film), or by the diffusion through the product layer, or by a surface chemical reaction. During leaching, the inert layer of solids shrinks toward the center. A porous film of the reaction product is formed around the inert core.

The slowest stage with the greatest resistance is the limiting step, and its intensification allows to increase the leaching efficiency.

The equations describing the several limiting stages of SCM [45] are given in Table 5. A large number of studies [40,42] show that the new version of SCM proposed by Dickenson and Heal [46] may be preferable to describe the kinetics of leaching reactions controlled by interfacial transfer and diffusion through the product layer (Equation (7) in Table 5).

Table 5. The shrinking core model (SCM) equations [45].

#	Limiting Step	Equation
1	Diffusion through the product layer (sp)	$1 - 3(1 - X)^{2/3} + 2(1 - X) = kt$
2	Diffusion through the product layer (pp)	$X^2 = kt$
3	Diffusion through the product layer (cp)	$X + (1 - X)\ln(1 - X) = kt$
4	Diffusion through the liquid film (sp)	$X = kt$
5	Surface chemical reactions (cp)	$1 - (1 - X)^{1/2} = kt$
6	Surface chemical reactions (sp)	$1 - (1 - X)^{1/3} = kt$
7	New shrinking core model	$1/3\ln(1 - X) + [(1 - X)^{-1/3} - 1] = kt$

sp—spherical particles, pp—plate particles, cp—cylinder particles, k—a chemical constant, X—the degree of iron recovery into the solution, and t—the leaching time.

According to Equation (7) in Table 5, if the interfacial transfer and diffusion through the product layer represent the limiting stage, then the function $1/3\ln(1 - X) + [(1 - X)^{-1/3} - 1]$ on the time "t" will be a straight line with the slope angle "k".

For the kinetic analysis using SCM, the equations for leaching of refractory gold-containing concentrate-1 at a 10% concentration of HNO_3, presented in Table 5, were evaluated. The obtained data made it possible to determine the correlation coefficient (R^2) showing the average square deviation of the experimental data from the straight line. The results of the calculations are shown in Table 6.

Table 6. SCM equations fitting.

#	SCM Equation	R^2			
		70 °C	75 °C	80 °C	85 °C
1	$1 - 3(1 - X)^{2/3} + 2(1 - X)$	0.976	0.935	0.922	0.954
2	X^2	0.962	0.905	0.883	0.918
3	$X + (1 - X)\ln(1 - X)$	0.973	0.928	0.913	0.946
4	X	0.559	0.334	0.259	0.348
5	$1 - (1 - X)^{1/2}$	0.624	0.436	0.374	0.477
6	$1 - (1 - X)^{1/3}$	0.645	0.461	0.410	0.517
7	$1/3\ln(1 - X) + [(1 - X)^{-1/3} - 1]$	0.990	0.969	0.963	0.984

As can be seen from the data obtained, SCM Equations (4)–(6) in Table 6 are poorly suited to describing these leaching reactions, since the correlation coefficient is less than 0.9. It is also obvious that the kinetic data best of all correspond to the new shrinking core model at all temperatures, which indicates diffusion limitations during leaching reactions.

The slope k for each straight line obtained by substituting the experimental data of leaching concentrate-1 with a 10% HNO_3 solution into the SCM Equation (Figure 7a) was calculated. Then Arrhenius plots for the dependence of ln$k$ on inverse temperature (Figure 7b) were used. Building a straight line $y = ax + b$ in this plane of coordinates made it possible to find the coefficient a, which determines the slope of straight line. Knowing the slopes in these coordinates, allowed to find the apparent activation energy of 60.276 kJ/mol, using Arrhenius law. According to the literature data [39,47], a high value of the activation energy is not always representative of the kinetic controlled reaction.

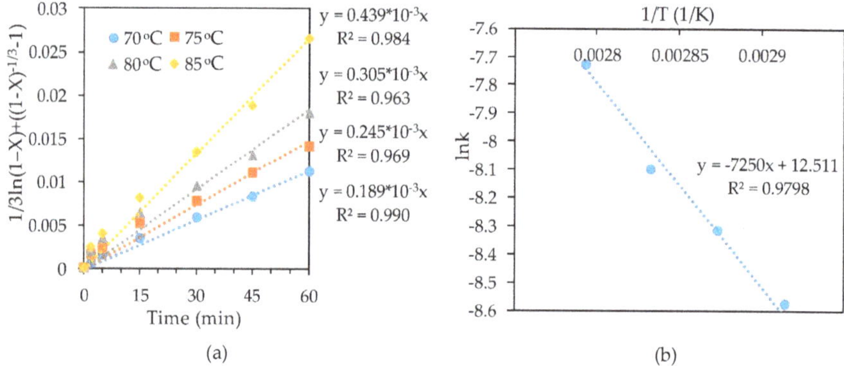

Figure 7. Calculation of slope k (**a**) and dependence of ln$k$-1/T (**b**) for leaching concentrate-1 with 10% HNO_3.

The slope of the straight lines (Figure 8a) obtained by substituting the concentrate-1 leaching results into the SCM equation at 70 °C at various concentrations of HNO_3 was plotted as ln$k$-ln(HNO_3) to determine the order with respect to nitric acid (Figure 8b). The resulting empirical order with respect to nitric acid concentration was 0.837.

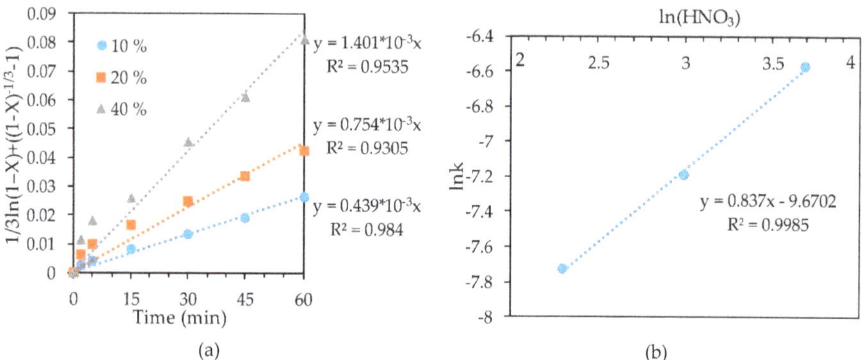

Figure 8. Calculation of slope k (**a**) and dependence ln$k$-ln(HNO_3) (**b**) for leaching conc-1 at 70 °C.

In the same way, we determined the empirical order with respect to sulfur content in the concentrate by plotting the dependence ln$k$-ln$S$ for leaching various concentrates with a 10% solution of nitric acid at 70 °C (Figure 9). The order with respect to sulfur was 2.948, which confirms the conclusions about the pronounced effect of the sulfide content in the concentrate on the degree of iron recovery.

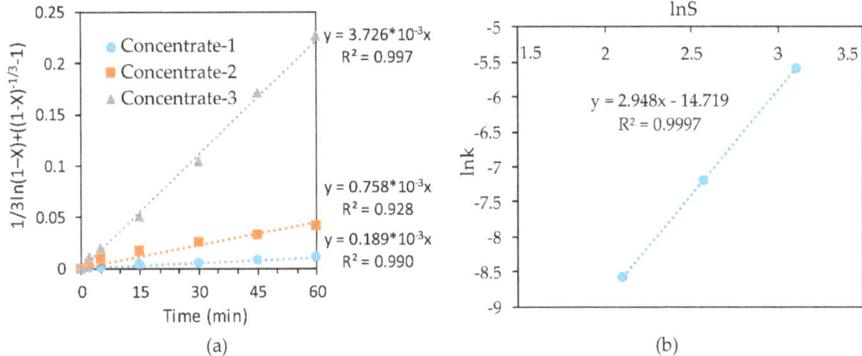

Figure 9. Calculation of slope k (**a**) and dependence of ln$k$-ln$S$ (**b**) for leaching of various concentrates at a temperature of 70 °C with a 10% HNO_3.

Substituting the Arrhenius equation (Equation (8)) into the equation of the new SCM model (Equation (7) in Table 5) gives Equation (9).

$$k = k_o e^{-E_a/RT}, \quad (8)$$

$$1/3\ln(1 - X) + [(1 - X)^{-1/3} - 1] = k_o e^{-E_a/RT} \cdot t. \quad (9)$$

In Equation (9), the Arrhenius constant k_o depends on the initial parameters of the process, including the initial sulfur content in the concentrate and the concentration of HNO_3 in the initial solution; hence, Equation (9) can be represented as follows (Equation (10)).

$$1/3\ln(1 - X) + [(1 - X)^{-1/3} - 1] = k_o(HNO_3)^n(S)^m e^{-E_a/RT} \cdot t, \quad (10)$$

where n and m are orders of concentration of HNO_3 and sulfur content in the original concentrate, respectively.

Based on the previously obtained results, the following equation for leaching refractory gold-containing concentrate with HNO_3 (Equation (11)) can be derived:

$$1/3\ln(1 - X) + [(1 - X)^{-1/3} - 1] = k_o(HNO_3)^{0.837}(S)^{2.948} e^{-60276/RT} \cdot t. \quad (11)$$

Building off the Arrhenius plots for all temperatures, HNO_3 concentrations and concentrates gives coefficients b of straight lines at a fixed slope a = 7250. The obtained values of the "b" coefficients and the corresponding correlation coefficients R^2 are shown in Table 7. To determine k_o, an exponent from the "a" coefficient was taken, and then it was divided by $(HNO_3)^{0.837}(S)^{2.948}$. The determined average k_o value was 87.811.

Table 7. Arrhenius constant determination.

S	HNO_3	a	R^2	k_o
8	10	12.51	0.980	85.917
8	20	12.99	0.995	86.173
8	40	13.66	0.998	84.973
13	10	13.97	0.972	88.673
13	20	14.63	0.999	95.291
13	40	15.12	0.999	87.348
22	10	15.61	0.986	96.532
22	20	16.11	0.964	88.930
22	40	16.54	0.995	76.464

Substituting this value into Equation (11) gives the following empirical equation describing the leaching process under study (Equation (12)):

$$1/3\ln(1 - X) + [(1 - X)^{-1/3} - 1] = 87.811(HNO_3)^{0.837}(S)^{2.948}e^{-60276/RT}\cdot t, \quad (12)$$

Table 8 shows the calculated correlation coefficient (R^2) of experimental data and data obtained using Equation (12). As can be seen from the table, the obtained empirical expression shows a high degree of convergence with experimental data at almost all temperatures, concentrations of HNO_3 and sulfur contents in the concentrate.

Table 8. Comparison of experimental data and the obtained empirical equation.

HNO_3	Concentrate-1				Concentrate-2				Concentrate-3			
	70	75	80	85	70	75	80	85	70	75	80	85
10%	0.988	0.997	0.995	0.984	0.927	0.940	0.940	0.956	0.996	0.963	0.964	0.972
20%	0.935	0.992	0.992	0.923	0.957	0.961	0.951	0.948	0.971	0.972	0.981	0.981
40%	0.949	0.995	0.996	0.996	0.960	0.956	0.960	0.960	0.959	0.941	0.930	0.918

4. Conclusions

The kinetics of dissolution of refractory sulfide gold-containing concentrates by a solution of HNO_3 in the temperature range of 70–85 °C was investigated. The data obtained allowed one to draw the following conclusions:

Increasing the temperature from 70 to 85 °C effected an increase in iron extraction from concentrate-1 from 35.53 to 48.52% after 1 h of leaching with a 10% solution of HNO_3. Increasing the concentration of HNO_3 from 10 to 40% has same effect. Changing sulfur content in the concentrate produces a much greater effect. The highest degree of iron recovery in 1 h for high-sulfur concentrate was 98.10%, while recovery from low-sulfur concentrate under the same conditions was 67.83%.

EDS and XRD showed elemental sulfur and non-leached arsenopyrite in the residue from leaching of low-sulfur concentrate, while quartz was the main phase in the residue from leaching of high-sulfur concentrate.

The iron recovery from the concentrate is well described by a new shrinking core model, which suggests that the speed of the process is controlled by interfacial diffusion and diffusion through the product layer. The calculated apparent activation energy was 60.276 kJ/mol, and the reaction order with respect to the initial concentration of HNO_3 was 0.837; the reaction order with respect to the initial S content in the concentrate was 2.948. The obtained data allowed us to derive a semi-empirical equation describing the leaching kinetics of iron:

$$1/3\ln(1 - X) + [(1 - X)^{-1/3} - 1] = 87.811(HNO_3)^{0.837}(S)^{2.948}e^{-60276/RT}\cdot t$$

Comparison of the experimental data obtained in the whole range of the studied parameters and the derived equation showed high convergence of the results. Thus, it can be concluded that the increase of sulfur content in the concentrate can be used to ensure more energy-efficient oxidation of sulfide minerals. The focus of our further research will be the study of HNO_3 regeneration and methods of arsenic disposal in the form of environmentally friendly compounds. The new study will aim at finding the optimal conditions for the process and could lay the basis for the development of an alternative commercial technology.

Author Contributions: Conceptualization, D.A.R.; methodology, K.A.K. and A.A.S.; validation, O.A.D., R.E.R.; formal analysis, K.A.K., D.A.R.; investigation, A.A.S. and O.A.D.; resources, R.E.R., D.A.R.; data curation, D.A.R.; writing—original draft preparation, A.A.S.; writing—review and editing, K.A.K., D.A.R.; visualization, R.E.R., O.A.D.; supervision, A.A.S.; project administration, D.A.R.; funding acquisition, D.A.R.

Funding: The research was funded by the Russian Science Foundation, grant number 18-19-00186. The SEM/EDS and microprobe analysis were funded by State Assignment, grant number 11.4797.2017/8.9.

Acknowledgments: Ekaterinburg Non-Ferrous Metal Processing Plant JSC are acknowledged for providing materials. Technicians at Ural Branch of Russian Academy of Sciences are acknowledged for their assistance with XRD, XRF, SEM, EDS, and ICP-MS analysis.

Conflicts of Interest: The authors declare no conflict of interest. The funders had no role in the design of the study; in the collection, analyses, or interpretation of data; in the writing of the manuscript, or in the decision to publish the results.

References

1. Genkin, A.D.; Wagner, F.E.; Krylova, T.L.; Tsepin, A.I. Gold-bearing arsenopyrite and its formation condition at the Olympiada and Veduga gold deposits (Yenisei Range, Siberia). *Geol. Ore Deposits* **2002**, *44*, 52–68.
2. Strižko, L.S. *Metallurgija zolota i sereba*; Misis: Moskva, Russia, 2001; ISBN 978-5-87623-083-6. (In Russian)
3. Vikentyev, I.V. Invisible and microscopic gold in pyrite: Methods and new data for massive sulfide ores of the Urals. *Geol. Ore Deposits* **2015**, *57*, 237–265. [CrossRef]
4. Marsden, J.; House, I. *The Chemistry of Gold Extraction*, 2nd ed.; Society for Mining, Metallurgy, and Exploration: Littleton, CO, USA, 2006; ISBN 978-0-87335-240-6.
5. Majzlan, J.; Chovan, M.; Andráš, P.; Newville, M.; Wiedenbeck, M. The nanoparticulate nature of invisible gold in arsenopyrite from Pezinok (Slovakia). *Neues JB. Miner. Abh.* **2010**, *187*, 1–9. [CrossRef]
6. Palenik, C.S.; Utsunomiya, S.; Reich, M.; Kesler, S.E.; Wang, L.; Ewing, R.C. "Invisible" gold revealed: Direct imaging of gold nanoparticles in a Carlin-type deposit. *Am. Mineral.* **2004**, *89*, 1359–1366. [CrossRef]
7. Yang, S.; Blum, N.; Rahders, E.; Zhang, Z. The nature of invisible gold in sulfides from the Xiangxi Au-Sb-W ore deposit in northwestern Hunan, People's Republic of China. *Can Mineral.* **1998**, *36*, 1361–1372.
8. Cabri, L.J.; Newville, M.; Gordon, R.A.; Crozier, E.D.; Sutton, S.R.; McMahon, G.; Jiang, D.-T. Chemical speciation of gold in arsenopyrite. *Can. Mineral.* **2000**, *38*, 1265–1281. [CrossRef]
9. Chen, T.T.; Cabri, L.J.; Dutrizac, J.E. Characterizing gold in refractory sulfide gold ores and residues. *JOM* **2002**, *54*, 20–22. [CrossRef]
10. Chryssoulis, S.L.; McMullen, J. Mineralogical investigation of gold ores. In *Developments in Mineral Processing*; Elsevier: Amsterdam, The Netherlands, 2005; Volume 15, pp. 21–71. ISBN 978-0-444-51730-2.
11. Nabojčenko, S.S. (Ed.) *Avtoklavnaja gidrometallurgija cvetnych metallov*; GOU UGTU-UPI: Ekaterinburg, Russia, 2002; ISBN 978-5-321-00065-6. (In Russian)
12. Simon, G.; Kesler, S.E.; Chryssoulis, S. Geochemistry and textures of gold-bearing arsenian pyrite, Twin Creeks, Nevada; implications for deposition of gold in carlin-type deposits. *Econ. Geol.* **1999**, *94*, 405–421. [CrossRef]
13. Fleet, M.E.; Mumin, A.H. Gold-bearing arsenian pyrite and marcasite and arsenopyrite from Carlin Trend gold deposits and laboratory synthesis. *Am. Mineral.* **1997**, *82*, 182–193. [CrossRef]
14. Tauson, V.L. Gold solubility in the common gold-bearing minerals: Experimental evaluation and application to pyrite. *Eur. J. mineral.* **1999**, *11*, 937–948. [CrossRef]
15. Zhou, Y.; Wang, K. Gold in the Jinya Carlin-type Deposit: Characterization and Implications. *JMMCE* **2003**, *2*, 83–100. [CrossRef]
16. Cabri, L.; Chryssoulis, S.L.; Villiers, J.; Laflamme, J.H.G.; Buseck, P.R. The nature of "invisible" gold in arsenopyrite. *Can. Mineral.* **1989**, *27*, 353–362.
17. Genkin, A.D.; Bortnikov, N.S.; Cabri, L.J.; Wagner, F.E.; Stanley, C.J.; Safonov, Y.G.; McMahon, G.; Friedl, J.; Kerzin, A.L.; Gamyanin, G.N. A multidisciplinary study of invisible gold in arsenopyrite from four mesothermal gold deposits in Siberia, Russian Federation. *Econ. Geol.* **1998**, *93*, 463–487. [CrossRef]
18. Liu, X.; Li, Q.; Zhang, Y.; Jiang, T.; Yang, Y.; Xu, B.; He, Y. Improving gold recovery from a refractory ore via Na2SO4 assisted roasting and alkaline Na2S leaching. *Hydrometallurgy* **2019**, *185*, 133–141. [CrossRef]
19. James, K.A.; Meliker, J.R.; Nriagu, J.O. Arsenic. In *International Encyclopedia of Public Health (Second Edition)*; Quah, S.R., Ed.; Academic Press: Oxford, UK, 2017; pp. 170–175. ISBN 978-0-12-803708-9.
20. Shoppert, A.; Loginova, I.; Rogozhnikov, D.; Karimov, K.; Chaikin, L. Increased As Adsorption on Maghemite-Containing Red Mud Prepared by the Alkali Fusion-Leaching Method. *Minerals* **2019**, *9*, 60. [CrossRef]
21. Safiur Rahman, M.; Khan, M.D.H.; Jolly, Y.N.; Kabir, J.; Akter, S.; Salam, A. Assessing risk to human health for heavy metal contamination through street dust in the Southeast Asian Megacity: Dhaka, Bangladesh. *Sci. Total Environ.* **2019**, *660*, 1610–1622. [CrossRef] [PubMed]

22. Quansah, R.; Armah, F.A.; Essumang, D.K.; Luginaah, I.; Clarke, E.; Marfoh, K.; Cobbina, S.J.; Nketiah-Amponsah, E.; Namujju, P.B.; Obiri, S.; et al. Association of Arsenic with Adverse Pregnancy Outcomes/Infant Mortality: A Systematic Review and Meta-Analysis. *Environ. Health Perspect.* **2015**, *123*, 412–421. [CrossRef]
23. Chung, J.-Y.; Yu, S.-D.; Hong, Y.-S. Environmental Source of Arsenic Exposure. *J. Prev. Med. Public Health* **2014**, *47*, 253–257. [CrossRef]
24. Ferrante, M.; Napoli, S.; Grasso, A.; Zuccarello, P.; Cristaldi, A.; Copat, C. Systematic review of arsenic in fresh seafood from the Mediterranean Sea and European Atlantic coasts: A health risk assessment. *Food Chem. Toxicol.* **2019**, *126*, 322–331. [CrossRef]
25. Arsenic. Available online: https://www.who.int/news-room/fact-sheets/detail/arsenic (accessed on 17 April 2019).
26. Sorg, T.J.; Chen, A.S.C.; Wang, L. Arsenic species in drinking water wells in the USA with high arsenic concentrations. *Water Res.* **2014**, *48*, 156–169. [CrossRef]
27. Adams, M.D. (Ed.) *Advances in Gold Ore Processing*, 1st ed.; Developments in mineral processing; Elsevier: Amsterdam, The Netherlands, 2005; ISBN 978-0-444-51730-2.
28. Paphane, B.D.; Nkoanet, B.B.M.; Oyetunjit, O.A. Kinetic studies on the leaching reactions in the autoclave circuit of the Tati Hydrometallurgical Demonstration Plant. *J. South. Afr. Inst. Min. Metall.* **2013**, *113*, 485–489.
29. Hourn, M. Refractory leaching solutions. *Aust. Min.* **2009**, *101*, 20.
30. Zaytsev, P.; Fomenko, I.; Chugaev, V.L.; Shneerson, M.Y. Pressure oxidation of double refractory raw materials in the presence of limestone. *Tsvetn. Met.* **2015**, 41–49. [CrossRef]
31. Dreisinger, D. Hydrometallurgical process development for complex ores and concentrates. *J. South. Afr. Inst. Min. Metall.* **2009**, *109*, 253–271.
32. Rogozhnikov, D.A.; Mamyachenkov, S.V.; Anisimova, O.S. Nitric Acid Leaching of Copper-Zinc Sulfide Middlings. *Metallurgist* **2016**, *60*, 229–233. [CrossRef]
33. Rogozhnikov, D.A.; Rusalev, R.E.; Dizer, O.A.; Naboychenko, S.S. Nitric acid loosening of rebellious sulphide concentrates containing precious metals. *Tsvetn. Met.* **2018**, *16*, 38–40. [CrossRef]
34. Anderson, C.G.; Harrison, K.D.; Krys, L.E. Theoretical considerations of sodium nitrite oxidation and fine grinding in refractory precious-metal concentrate pressure leaching. *Miner. Metall. Proc.* **1996**, *13*, 4–11. [CrossRef]
35. Van Weert, G.; Fair, K.J.; Schneider, J.C. Prochem's NITROX Process. *CIM Bull.* **1986**, *79*, 84–85.
36. Beattie, M.J.V.; Ismay, A. Applying the redox process to arsenical concentrates. *JOM* **1990**, *42*, 31–35. [CrossRef]
37. La Brooy, S.R.; Linge, H.G.; Walker, G.S. Review of gold extraction from ores. *Miner. Eng.* **1994**, *7*, 1213–1241. [CrossRef]
38. Gao, G.; Li, D.; Zhou, Y.; Sun, X.; Sun, W. Kinetics of high-sulphur and high-arsenic refractory gold concentrate oxidation by dilute nitric acid under mild conditions. *Miner. Eng.* **2009**, *22*, 111–115. [CrossRef]
39. Gok, O.; Anderson, C.G.; Cicekli, G.; Cocen, E.I. Leaching kinetics of copper from chalcopyrite concentrate in nitrous-sulfuric acid. *Physicochem. Probl. Mi.* **2014**, *50*, 399–413.
40. Huang, Y.; Dou, Z.; Zhang, T.; Liu, J. Leaching kinetics of rare earth elements and fluoride from mixed rare earth concentrate after roasting with calcium hydroxide and sodium hydroxide. *Hydrometallurgy* **2017**, *173*, 15–21. [CrossRef]
41. Kocan, F.; Hicsonmez, U. Leaching kinetics of celestite in nitric acid solutions. *Int. J. Min. Met. Mater.* **2019**, *26*, 11–20. [CrossRef]
42. Li, K.; Chen, J.; Zou, D.; Liu, T.; Li, D. Kinetics of nitric acid leaching of cerium from oxidation roasted Baotou mixed rare earth concentrate. *J. Rare Earth.* **2019**, *37*, 198–204. [CrossRef]
43. Fajaryanto, R.; Nurqomariah, A. Silvia Acid leaching and kinetics study of cobalt recovery from spent lithium-ion batteries with nitric acid. In Proceedings of the 3rd International Tropical Renewable Energy Conference "Sustainable Development of Tropical Renewable Energy" (i-TREC 2018), Bali, Indonesia, 6–8 September 2018; Volume 67. [CrossRef]
44. Anderson, C.G.; Twidwell, L.G. Hydrometallurgical processing of gold-bearing copper enargite concentrates. *Can. Metall. Quart.* **2008**, *47*, 337–346. [CrossRef]
45. Levenspiel, O. *Chemical Reaction Engineering*, 3rd ed.; Wiley: New York, NY, USA, 1999; ISBN 978-0-471-25424-9.

46. Dickinson, C.F.; Heal, G.R. Solid-liquid diffusion controlled rate equations. *Thermochim. Acta* **1999**, *340–341*, 89–103. [CrossRef]
47. Baba, A.A.; Adekola, F.A. Hydrometallurgical processing of a Nigerian sphalerite in hydrochloric acid: Characterization and dissolution kinetics. *Hydrometallurgy* **2010**, *101*, 69–75. [CrossRef]

Article

Enhanced Desilication of High Alumina Fly Ash by Combining Physical and Chemical Activation

Yanbing Gong [1,2], Junmin Sun [2], Shu-Ying Sun [3], Guozhi Lu [1] and Ting-An Zhang [1,*]

1 Key Laboratory of Ecological Metallurgy of Multi-metal Intergrown Ores of Ministry of Education, Special Metallurgy and Process Engineering Institute, Northeastern University, Shenyang 110819, China; gongyb2001@163.com (Y.G.); lvgz@smm.neu.edu.cn (G.L.)

2 High-alumina Coal Resources Development and Utilization R&D Center, Datang International Power Generation Co., Ltd., Hohhot 010050, China; sunjmdt@163.com

3 National Engineering Research Center for Integrated Utilization of Salt Lake Resource, East China University of Science and Technology, Shanghai 200237, China; shysun@ecust.edu.cn

* Correspondence: zta2000@163.net; Tel./Fax: +86-024-8368-1563

Received: 14 March 2019; Accepted: 2 April 2019; Published: 4 April 2019

Abstract: In this work, a physical–chemical activation desilication process was proposed to extract silica from high alumina fly ash (HAFA). The effects of fly ash size, hydrochloric acid concentration, acid activation time, and reaction temperature on the desilication efficiency were investigated comprehensively. The phase and morphology of the original fly ash and desilicated fly ash were analyzed by X-ray diffraction (XRD) and scanning electron microscopy–energy-dispersive X-ray spectroscopy (SEM-EDS). Compared with the traditional desilication process, the physical–chemical activation desilication efficiency is further increased from 38.4% to 53.2% under the optimal conditions. Additionally, the kinetic rules and equations were confirmed by the experimental data fitting with shrinking core model of liquid–solid multiphase reaction. Kinetic studies show that the enhanced desilication process is divided into two processes, and both steps of the two-step reaction is controlled by chemical reaction, and the earlier stage activation energy is 52.05 kJ/mol and the later stage activation energy is 58.45 kJ/mol. The results of mechanism analysis show that physical activation breaks the link between the crystalline phase and the amorphous phase, and then a small amount of alkali-soluble alumina in the amorphous phase is removed by acid activation, thereby suppressing the generation of side reactions of the zeolite phase.

Keywords: high alumina fly ash; desilication rate; physical–chemical activation; alumina silica mass ratio; kinetics

1. Introduction

High alumina fly ash (HAFA) is a byproduct of high-temperature combustion of high-alumina coal in thermal power plants. The annual emission of HAFA is approximately 30 million tons [1]. Except for a small amount that is utilized, most of the HAFA is currently stockpiled in northwestern China, not only occupying farmland, but also giving rise to severe pollution of water, atmosphere, and soil [2,3]. On the other hand, the high alumina content (approximately 50%) in high alumina fly ash makes it a valuable recycling resource. Furthermore, HAFA contains critical metals such as gallium [4], lithium [5], and rare earths [6,7] that can be extracted in the alumina extraction process as byproducts. Many extraction techniques, such as predesilication soda lime sintering [8], limestone sintering [9], acid leaching [10,11], ammonium sulfate [12] and submolten salt methods [13], have been developed to extract alumina from HAFA. Only the predesilication soda lime sintering process was carried out at an industrial scale (200 kt per year) by China Datang Corporation in the province of

Inner Mongolia [14]. This formed a circular economy industry (shown in Figure 1) in Tuoketou County, Inner Mongolia. The recycling route of thermal power plant HAFA is presented in Figure 1.

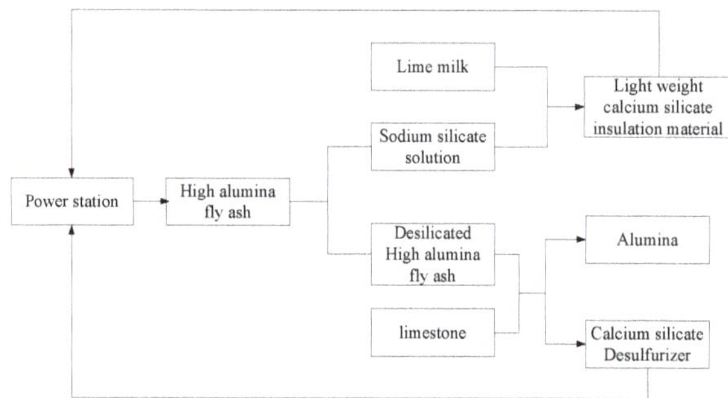

Figure 1. Recycling route of thermal power plant high alumina fly ash (HAFA).

Here, a traditional alkali predesilication method was used to obtain desilicated high alumina fly ash (DHAFA) and sodium silicate solution. The main problem for this technology is that the Al_2O_3/SiO_2 mass ratio (A/S) is only elevated to 1.7–1.9. Not only does this make the desilication rate lower, it also increases the difficulty of extraction of alumina by the soda lime sintering process [15]. Because of the low A/S of DHAFA, the main reaction equipment rotary kiln often scabs the inner wall, forming large clinker eggs, and forming hard clinker [16], giving rise to the consequent decrease of the alumina extraction rate, damage to equipment, and increased production cost. Therefore, it is necessary to improve the desilication rate.

$$2NaOH + SiO_2 \text{ (amorphous)} = Na_2SiO_3 + H_2O \text{ (main reaction)} \quad (1)$$

$$10Na_2SiO_3 + 3Al_2O_3 \text{ (amorphous)} + 19H_2O = Na_6Al_6Si_{10}O_{32}\cdot 12H_2O\downarrow + 14NaOH \text{ (side reaction)} \quad (2)$$

Many researchers have reported that the amorphous SiO_2 in the high alumina fly ash can be selectively dissolved in an NaOH solution [17], thereby leading to a decrease in the amount of silicon–calcium slag generated during the subsequent extraction process of Al_2O_3 and achieving the efficient recovery of the silicon resource [18,19]. These studies also indicated that the final products were significantly affected by the reaction conditions [20,21]. The mechanism of traditional predesilication can be clearly explained as follows: the amorphous SiO_2 in the HAFA dissolved in an NaOH solution and formed sodium silicate as the main reaction (Equation (1)), and meanwhile, a small amount of Al_2O_3 was dissolved in an NaOH solution and formed sodium aluminate, and then sodium silicate reacted with sodium aluminate and formed zeolite as a side reaction (Equation (2)) that inhibits the desilication rate. Zhang et al. [22] has investigated a new process to improve the desilication rate of HAFA. The process is divided into three steps; traditional desilication, acid activation, and secondary desilication. Though the A/S of HAFA is raised from 1.2 to 2.85, the acid activation process consumes a large amount of acid to neutralize the alkali of zeolite generated in reaction Equation (2), so it is uneconomical. A desilication process after 12-h alkali pretreatment [23] was used to improve the A/S of DHAFA from 1.92 to 2.51, and the mechanism can be explained as the zeolite P generation in the pretreatment process blocking the formation of hydroxysodalite in the desilication process. It was found that the desilicated slurry can easily settle at the bottom of the tank and is difficult to filter after a long time of stirring in actual industrial production. A mechanism–chemical synergistic activation desilication process is used to prepare mullite ceramic materials [24]. The process uses harsh reaction conditions to remove the impurities in HAFA. The mechanism [25] has been analyzed. However, the specific effect of the reaction conditions and kinetics of desilication process were not described.

Based on the previous work [21–23], in this study, a novel process based on physical and chemical activation was developed in order to inhibit the side reaction during desilication and to increase the desilication efficiency. This process is moderate, controllable, and easy to transfer to industrial use. Moreover, the Al_2O_3/SiO_2 ratio in desilication fly ash was significantly improved, which is beneficial to the subsequent Al_2O_3 extraction process. The effect of the activation reaction conditions on the desilication efficiency and its mechanism were investigated, and the kinetic rules and equations of the SiO_2 leaching process were confirmed.

2. Materials and Methods

2.1. Materials

Hydrochloric acid (Beijing Chemical Works, Beijing, China, 36–38 wt. %) and NaOH (Sinopharm Chemical Reagent Co., Ltd., Shanghai, China, ≥96 wt. %) used in this study were of the analytical grade. HAFA was generated by a coal-fired boiler of a power plant of the Inner Mongolia Datang International Tuoketuo Power Generation Co., Ltd., Hohhot, China.

Table 1 summarizes the chemical composition of the original fly ash. The metal component contents were evaluated by chemical analysis methods. Table 1 shows that Al_2O_3 and SiO_2 were the main components of the fly ash. Small amounts of iron oxide and calcium oxide are also present. Thus, this kind of fly ash had a high value for aluminum and silicon production. Additionally, the original fly ash contains a certain amount of rare metals Ga and Li.

Table 1. Chemical composition of HAFA.

Composition	Al_2O_3	SiO_2	Fe_2O_3	CaO	Na_2O	Li	Ga	A/S
Content (wt. %)	51.38	37.97	2.01	2.00	0.38	0.0384	0.0076	1.35

2.2. Processes and Methods

The physical–chemical activation desilication process includes three steps: "physical activation", "chemical activation", and "alkali predesilication".

During physical activation, the HAFA was milled for different times (5, 15, 30, 45, and 60 s) in a circular disk milling machine with a filling rate of approximately 60%.

During chemical activation, the milling HAFA was activated by hydrochloric acid with different acid concentration (4%, 6%, 8%, and 10%), reaction temperatures (30, 60, and 90 °C), and reaction times (0.5, 1, 1.5, 2, and 2.5 h) under a determined *L/S* ratio (equal to 4). The slurries were then filtered to obtain solid pretreatment HAFA (PHAFA).

During alkali predesilication, the original fly ash or physical–chemical activated solid was mixed with NaOH solutions with different alkali concentrations (Na_2O concentrations of 100, 120, 140, 160 g/L), reaction times (10, 20, 30, 60, 90, 120, 150, and 180 min) under the determined conditions (T = 95 °C, L/S = 3.5) during the alkali predesilication. The slurries were filtered to obtain DHAFA or intensified DHAFA (IDHAFA).

The silica leaching ratio was calculated as follows:

$$\eta SiO_2 = \frac{WD - W_1D_1}{WD} \times 100\% \quad (3)$$

where η is the desilication rate, W is the mass of HAFA, W_1 is the mass of DHAFA, D is the percentage content of SiO_2 in HAFA, and D_1 is the percentage content of SiO_2 in DHAFA or IDHAFA.

2.3. Characterization

The chemical composition of HAFA was analyzed by chemical analysis methods. Particle size distribution of HAFA was evaluated using a Sympatec GmbH particle size analyzer (OASIS/Dry,

Sympatec GmbH, Clausthal-Zellerfeld, Germany). The surface structure and the distribution of the elements were observed by scanning electron microscopy (JEOL5800SV, JEOL, Tokyo, Japan). The phase composition of HAFA was evaluated by X-ray diffraction (X' Pert PRO MPD, PANalytical B.V., Almelo, The Netherlands). The specific surface area and pore size analysis were performed using a Nova4000e high-speed surface tester (Nova4000e, Quantachrome, Boynton Beach, FL, USA). The molybdenum blue colorimetric method can be used to examine the content of SiO_2 in the filtrate.

3. Results and Discussion

3.1. *Effects of Physical Activation on the Desilication Rate*

Figure 2 shows the effect of physical activation time on the desilication rate, and A/S and Na_2O content of IDHAFA under acid activation conditions (T = 90 °C, acid concentration 6%, L/S = 4 and t = 1 h) and alkali predesilication conditions (T = 95 °C, C = 120 g/L, L/S = 3, and t = 2.5 h). It is observed that the desilication rate of HAFA is as high as 47% only due to the chemical activation. However, the desilication rate of HAFA is further improved to 51.9% after combined physical and chemical activation. Moreover, the alkali content of IDHAFA is decreased, indicating that the side reaction of the desilication process is suppressed, thereby improving the desilication rate. Considering the cost and effect of physical activation data presented in Table 2, the optimal activation time is 15 s.

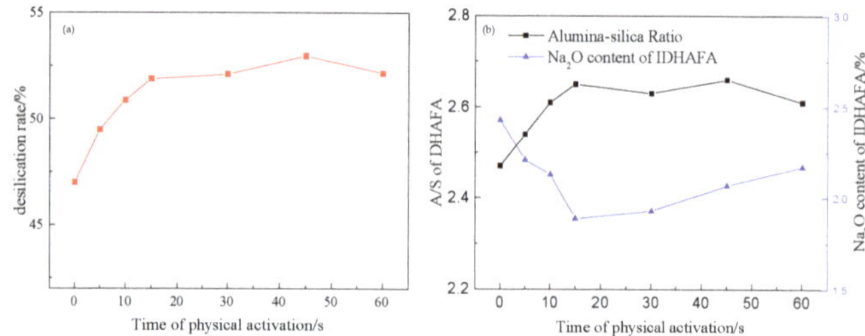

Figure 2. Effect of physical activation time: (**a**) desilication rate, (**b**) A/S and Na_2O content of IDHAFA.

Table 2. Effect of physical activation time on the particle size of HAFA (μm).

Time/s	0	5	10	15	20	25	30	45	60
D50/μm	23.03	5.22	4.86	4.36	3.37	3.12	2.84	2.02	2.01

3.2. *Effects of Chemical Activation on the Desilication Rate*

The data for the effects of acid concentration, reaction temperature, and reaction time on the desilication rate, A/S of DHAFA and Na_2O content of IDHAFA for physical activation time of 15 s, and alkali predesilication conditions (T = 95 °C, C = 120 g/L, L/S = 3, and t = 2.5 h) are shown in Figures 3 and 4.

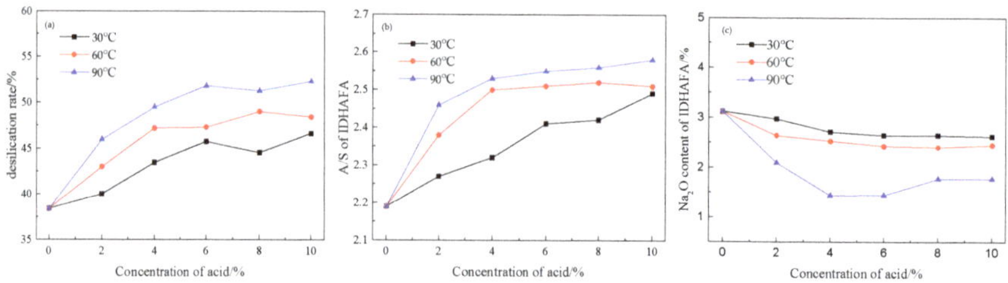

Figure 3. Effect of concentration of acid: (**a**) desilication rate, (**b**) A/S of intensified desilicated high alumina fly ash (IDHAFA), (**c**) Na_2O content of IDHAFA.

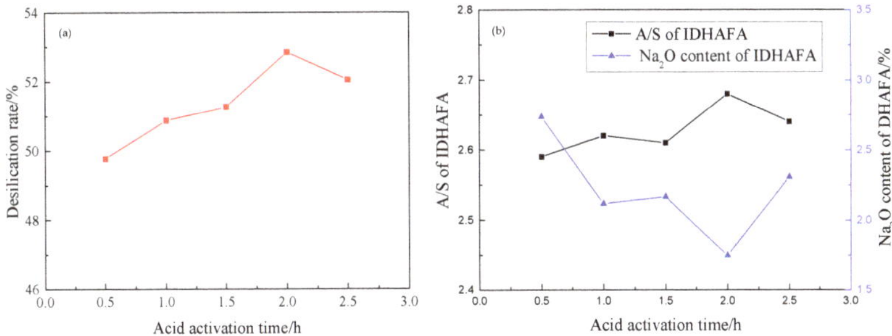

Figure 4. Effect of acid activation time (T = 95 °C): (**a**) desilication rate, (**b**) A/S and Na_2O of IDHAFA.

It is observed from Figure 3a that the desilication rate of the HAFA after acid activation is greatly improved compared with the conventional predesilication. The desilication rate is also improved with increasing temperature. The desilication rate is enhanced by more than 50% when the temperature is 90 °C. In addition, an increase in the acid concentration also enhances the desilication rate. It is observed from Figure 3b,c that the variation of the IDHAFA aluminum: silicon ratio is completely consistent with the desilication rate, but the change of the sodium oxide content in the IDHAFA shows a completely opposite trend to that of the desilication rate. This shows that the desilication fly ash aluminum: silicon ratio can reflect the desilication rate, while the sodium oxide content indicates the degree of the side reaction, which is inversely proportional to the desilication rate.

Figure 4a shows that the effect of the acid activation time on the desilication rate is also relatively large. In the range of 0–2 h, the desilication rate increases strongly with increased acid activation time. Flocculent silica gel was produced in the chemical activation process after 2 h, which makes the slurry becomes very viscous and difficult to filter, and the filter cake contains a large amount of acid, resulting in lower efficiency in desilication. It is observed from Figure 4b that the change in the desilication rate is consistent with the change in the ratio of IDHAFA to aluminum and silicon, and at the same time with the desilication powder. The change in the content of sodium oxide in IDHAFA is inversely proportional. Based on these experiments, it can be concluded that the optimal conditions for acid activation are: temperature of 90 °C, acid concentration of 6%, liquid: solid ratio of 4:1, and acid activation reaction time of 2 h. After chemical activation under these conditions, the metal ions are partially leached into the solution: and the dissolution rates of iron oxide and calcium oxide were 37.4% and 38.2%, respectively. Meanwhile, Al_2O_3 reached a dissolution rate of approximately 3.0%, but SiO_2 is hardly dissolved during the hydrochloric acid activation.

3.3. Effects of Alkali Predesilication on the Desilication Rate

Figures 5 and 6 show the alkali concentration, reaction time, and temperature effects on the direct desilication of the original fly ash and desilication after activation for the physical activation time of 15 s, and acid activation conditions (T = 90 °C, acid concentration 6%, L/S = 4 and t = 2 h).

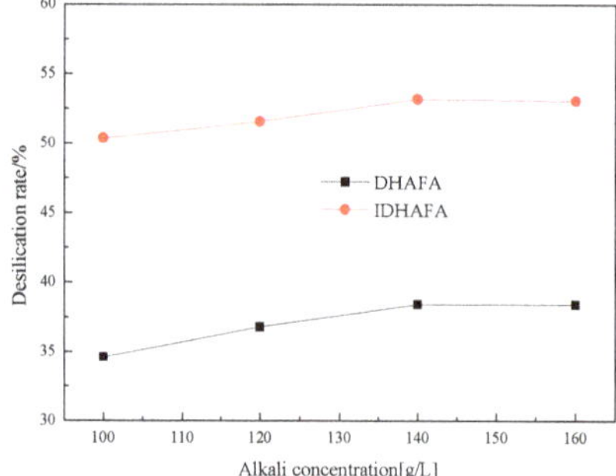

Figure 5. Effect of alkali concentration on the desilication rate of IDHAFA and DHAFA (Predesilication conditions: $T = 95$ °C, $L/S = 3.5$, and $t = 150$ min).

It can be concluded from Figure 5 that the improvement of desilication efficiency after enhanced desilication is obvious. Generally speaking, increasing the concentration of sodium hydroxide is beneficial to the progress of reaction Equation (1). However, a high concentration of sodium hydroxide will also promote the progress of reaction Equation (2), and this is detrimental to desilication. According to the results in Figure 5, $C(Na_2O) = 140$ g/L was required for the maximal desilication rate.

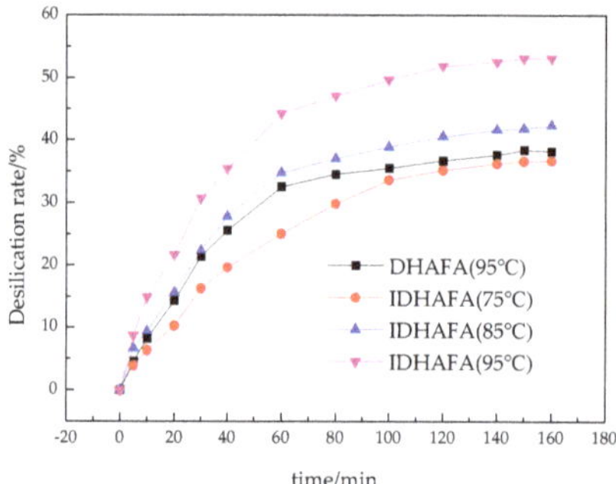

Figure 6. Effect of reaction time and temperature on the desilication rate of IDHAFA and DHAFA (Predesilication conditions: $C(Na_2O) = 140$ g/L, and $L/S = 3.5$).

It can be seen from Figure 6 that the effect of temperature on the desilication efficiency is very large. Although the enhanced desilication rate is much higher than the direct desilication at 95 °C, the desilication rate is greatly reduced at 75 °C. In the 0–160 min period, the desilication rate increases with the increase of reaction time. However, as the reaction time increases, the reaction rate continues to decrease. The change of reaction rate divides the desilication reaction into two stages.

The data shown in Figures 5 and 6 clearly illustrate the great difference between the two desilication methods. Compared with direct desilication, the desilication rate after activation is greatly improved from 38.4% to 53.2% under the predesilication conditions ($T = 95$ °C, $C(Na_2O) = 140$ g/L, $L/S = 3.5$, and $t = 2.5$ h).

3.4. Kinetics of the Desilication Process

As already mentioned, the reaction Equation (1) is the main reaction of the desilication reaction process. In order to examine the SiO_2 leaching kinetics of the enhanced desilication process, a shrinking core model of a liquid–solid multiphase reaction was used to fit the SiO_2 leaching kinetics data. According to the reaction rate changing node in Figure 6, the SiO_2 leaching process is divided into two stages, and the fitting data are shown in Figure 7.

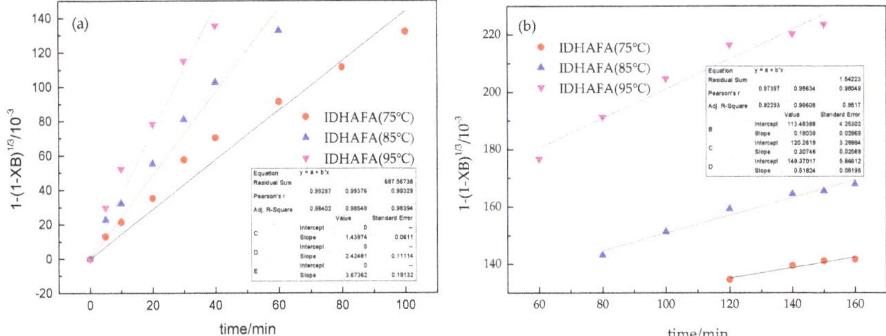

Figure 7. Mathematical fitting of data of SiO_2 leaching kinetics in enhanced desilication process: (**a**) The earlier stage, (**b**) The later stage.

There are two stages in the traditional desilication reaction process; the earlier stage is limited by the surface reaction, while in the later stage, the internal diffusion on the solid product layer is the rate-controlling step [26]. But, the physical–chemical activation desilication is significantly different. As exhibited in Figure 7a,b, the desilication process was controlled by chemical reaction in 0–160 min, in the temperature range of 75–95 °C, and both stages are controlled by chemical reaction. Under experimental conditions, the kinetic equation can be written as

$$1 - (1 - X_B)^{1/3} = k_1 t \quad (4)$$

$$1 - (1 - X_B)^{1/3} = k_2 t \quad (5)$$

where X_B is the desilication rate, t is the reaction time, k_1 is the rate constant of the earlier stage, k_2 is the rate constant of the later stage.

According to the Arrhenius equation $\ln k = \ln A - Ea/RT$, Plot $\ln k$ to $1/T$, as exhibited in Figure 8, both of the two stages have a good linear relationship. Calculated based on the slope of the fitted curve, the earlier stage's activation energy is 52.05 kJ/mol, the later stage's activation energy is 58.45 kJ/mol.

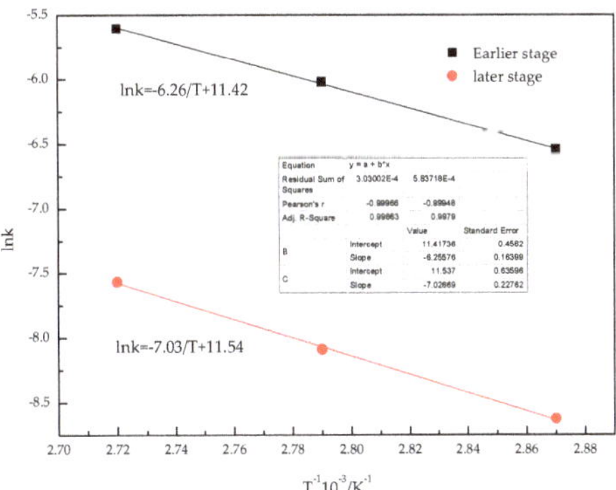

Figure 8. Arrhenius plot of SiO_2 leaching kinetics in the enhanced desilication process.

In the traditional desilication process, as the reaction time extends, the formation of zeolite in the side reaction increases, and a film is gradually formed in the fly ash particles. Therefore, the reaction is controlled by chemical reaction in the earlier stage and by diffusion in the later stage. But, the amount of zeolite is greatly reduced by enhanced desilication, so diffusion control is eliminated, and the enhanced desilication process is only controlled by the chemical reaction, which indicates that the set-up of the previous kinetic model is correct.

3.5. Mechanism

As shown in Figure 9, the phases of PHAFA and HAFA are basically the same, indicating that the physical–chemical activation process does not change the phase composition of HAFA. Moreover, according to the previous experimental results, the treatment only removed part of the iron oxide, calcium oxide, and a small amount of alumina; however, mullite, corundum, quartz, and glass-phase silica did not react, so there is no phase change before and after treatment. The bulge between 15° and 25° in the XRD patterns of both DHAFA and IDHAFA disappears because the activated amorphous SiO_2 was dissolved by the alkali solution during desilication, and the new zeolite phase appears in DHAFA because the soluble alumina enters the desilication solution and reacts with sodium silicate. It is also observed that there are few zeolite peaks in IDHAFA because the amount of soluble alumina decreases during the activation process, weakening the spontaneous side reaction (zeolite).

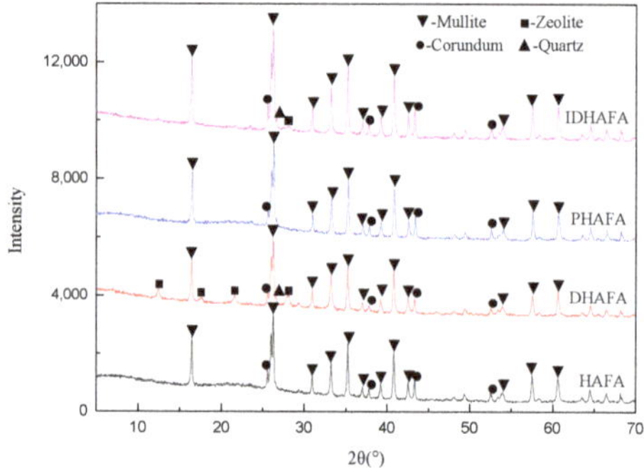

Figure 9. X-ray diffraction (XRD) patterns of HAFA, DHAFA, PHAFA, and IDHAFA.

Figure 10 shows the obtained N_2 adsorption isotherms and pore size distribution. It is observed that the specific surface areas are in the order of DHAFA > IDHAFA > PHAFA > HAFA. Very low N_2 adsorption is observed in HAFA (surface area = 2.886 m^2/g), but a high N_2 adsorption is observed for PHAFA for relative pressure (P/P0) between 0.5–1.0 (surface area = 13.737 m^2/g), IDHAFA (surface area = 18.519 m^2/g), and DHAFA (surface area = 21.460 m^2/g). The larger specific surface area of PHAFA than HAFA is ascribed to the removal of iron oxide and calcium oxide after physical and chemical activation. The increased surface area of IDHAFA and DHAFA may be related to the removal of amorphous silica wrapped in the mullite and the formation of a new zeolite phase. It also indicated that there are fewer mesopores in HAFA (4 nm) and PHAFA (3 nm); however, except for the smaller mesopores (3–4 nm) of DHAFA and IDHAFA, larger mesopores (13 nm) appeared in DHAFA, and even larger mesopores (10–14 nm) appeared in IDHAFA.

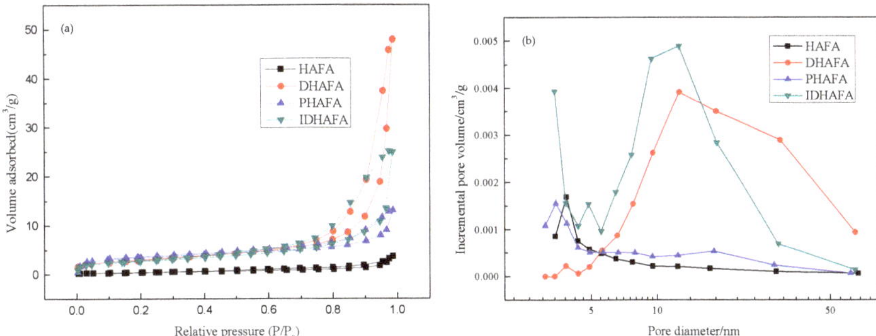

Figure 10. Nitrogen adsorption–desorption isotherms and pore size distribution of HAFA, DHAFA, PHAFA, and IDHAFA. (**a**) Nitrogen adsorption–desorption isotherms, (**b**) Pore size distribution.

Figure 11 shows the weight loss curve under conditions of protective gas air atmosphere and heating rate of 10 °C/min. It is observed that both HAFA and PHAFA have smaller weight loss because there are small amounts of unburned carbon removed. On the other hand, DHAFA and IDHAFA show higher weight losses due to the formation of byproduct zeolite containing crystalline water. When the temperature is higher than 700 °C, no weight loss was observed. It should be noted that the weight loss of IDHAFA is still less than that of DHAFA, which also indicates that the pretreatment suppresses the occurrence of side reactions of fly ash after desilication and reduces the formation of zeolite.

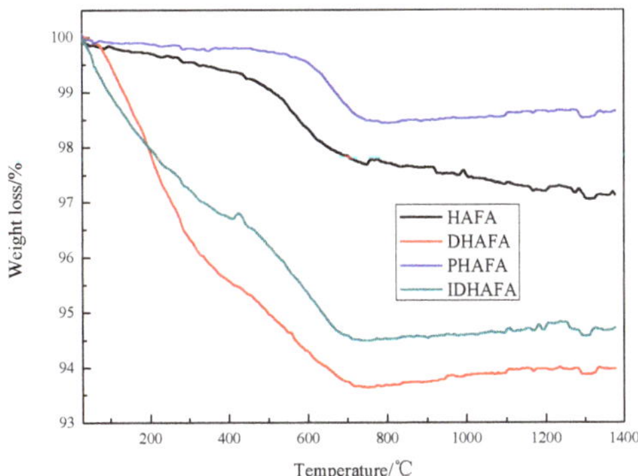

Figure 11. Thermogravimetric (TG) results for HAFA, DHAFA, PHAFA, and IDHAFA.

As demonstrated in Figure 12, the shape of the raw HAFA (Figure 12a) is more complicated. The spherical mullite and the irregular glass are intertwined and inlaid with each other, and the interface is not clear. After the conventional predesilication reaction, it is found that DHAFA (Figure 12c) is mainly composed of spherical mullite crystals, and the irregular glass phase essentially disappears. However, the spherical shape from the EDS analysis indicates that a large amount of the zeolite particles is covered, indicating a small amount of soluble oxidation in the conventional predesilication reaction. The aluminum reacted with sodium silicate solution re-enters DHAFA, resulting in a low desilication rate. After the physical and chemical activation treatment, it is observed that the mullite and the amorphous glass phase of PHAFA (Figure 12b) have been separated. This is highly beneficial for desilication. The surface of IDHAFA (Figure 12e) is smoother than the DHAFA (Figure 12c), and the particles in IDHAFA mainly consist of pure mullite crystals. Furthermore, the EDS analysis showed that the alkalinity of IDHAFA (Na 0.9 wt. %) (Figure 12f) is less than DHAFA (Na 2.39 wt. %) (Figure 12d), indicating that the side reaction of the zeolite ($Na_6Al_6Si_{10}O_{32}\cdot 12H_2O$) was greatly reduced.

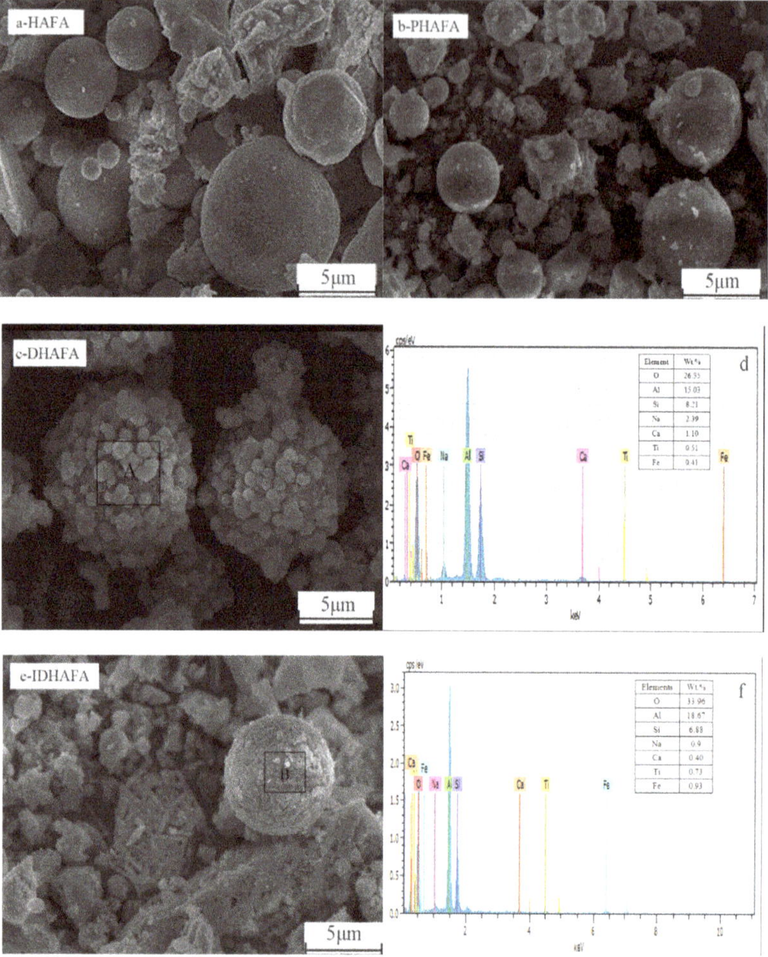

Figure 12. Scanning electron microscopy (SEM) and energy-dispersive X-ray spectroscopy (EDS) results: (**a**) HAFA, (**b**) PHAFA, (**c**,**d**) DHAFA, (**e**,**f**) IDHAFA.

As a conclusion, the mechanism of enhanced desilication can be summarized as follows.

First, the spherical mullite and the amorphous glass phase are separated by physical activation, and the reactivity of the SiO_2 and Al_2O_3 in the glass phase is improved.

Second, the Al_2O_3 in the glass phase is dissolved and removed by hydrochloric acid.

$$Al_2O_3\ (\text{amorphous}) + HCl = AlCl_3 + H_2O \quad (6)$$

Third, the SiO_2 in the glass phase is dissolved by sodium hydroxide with almost no side reactions to form zeolite ($Na_6Al_6Si_{10}O_{32}\cdot 12H_2O$). Therefore, only the mullite, corundum, and quartz are left in the fly ash after enhanced desilication.

$$2NaOH + SiO_2\ (\text{amorphous}) = Na_2SiO_3 + H_2O \quad (7)$$

4. Conclusions

Improvement in desilication efficiency is important not only for obtaining more silicon resources, but also for more economical and efficient extraction of alumina.

A new process that combined physical and chemical activation and predesilication has been proposed. By using physical and chemical activation, the desilication efficiency of HAFA increased from 38.4% to 53.2%. The reaction temperature has a significant effect on the desilication efficiency.

Kinetic study results show that the enhanced desilication process after physical and chemical activation fits with an unreacted shrinking core model of the liquid–solid reaction. The desilication reaction is divided into two stages. The earlier stage is controlled by chemical reaction and the activation energy is 52.05 kJ/mol and the later stage is also controlled by chemical reaction and the activation energy is 58.45 kJ/mol.

This result indicates that physical and chemical activation suppresses the formation of zeolite, thereby improving the desilication efficiency and further improving the *A/S* of the fly ash; results which are very advantageous for the next step of alumina extraction.

Author Contributions: Methodology, J.S.; experiments, data analyze and writing—original draft preparation, Y.G.; writing—review and editing S.-Y.S. and T.-A.Z.; formal analysis, G.L.

Acknowledgments: This research was funded by National Key R& D Program of China grant number 2017YFB0603103.

Conflicts of Interest: The authors declare no conflict of interest.

References

1. Sun, J.; Chen, P. Resourcing utilization of high alumina fly ash. *Adv. Mater.* **2013**, *652–654*, 2570–2575. [CrossRef]
2. Blissett, R.S.; Rowson, N.A. A review of the multi-component utilisation of coal fly ash. *Fuel* **2012**, *97*, 1–23. [CrossRef]
3. Jing, L.; Dong, Y.; Dong, X.; Hampshire, S.; Li, Z.; Zhu, Z.; Li, L. Feasible recycling of industrial waste coal fly ash for preparation of anorthite-cordierite based porous ceramic membrane supports with addition of dolomite. *J. Eur. Ceram. Soc.* **2016**, *36*, 1059–1071.
4. Li, S.; Wu, W.; Li, H.; Hou, X. The direct adsorption of low concentration gallium from fly ash. *Sep. Technol.* **2016**, *51*, 395–402. [CrossRef]
5. Hu, P.; Hou, X.; Zhang, J.; Li, S.; Hao, W.; Damø, A.J.; Li, H.; Wu, Q.; Xi, X. Distribution and occurrence of lithium in high-alumina-coal fly ash. *Int. J. Coal. Geol.* **2018**, *189*, 27–34. [CrossRef]
6. Dai, S.; Lei, Z.; Peng, S.; Chou, C.L.; Wang, X.; Yong, Z.; Dan, L.; Sun, Y. Abundances and distribution of minerals and elements in high-alumina coal fly ash from the jungar power plant, Inner Mongolia, China. *Int. J. Coal. Geol.* **2010**, *81*, 320–332. [CrossRef]
7. Lanzerstorfer, C. Fly ash from coal combustion: Dependence of the concentration of various elements on the particle size. *Fuel* **2018**, *228*, 263–271. [CrossRef]
8. Bai, G.; Teng, W.; Wang, X.; Qin, J.; Xu, P. Alkali desilicated coal fly ash as substitute of bauxite in lime-soda sintering process for aluminum production. *Trans. Nonferr. Met. Soc.* **2010**, *20*, s169–s175. [CrossRef]
9. Xiao, H.W.; Bao, D.W.; Yong, F.X.; Xiao, T.L.; Qi, S. The optimization of sintering process for alumina extraction from fly ash. *Adv. Mater.* **2014**, *878*, 264–270.
10. Cheng-You, W.U.; Hong-Fa, Y.U.; Zhang, H.F. Extraction of aluminum by pressure acid-leaching method from coal fly ash. *Trans. Nonferr. Met. Soc.* **2012**, *22*, 2282–2288.
11. Shemi, A.; Ndlovu, S.; Sibanda, V.; Dyk, L.D.V. Extraction of alumina from coal fly ash using an acid leach-sinter-acid leach technique. *Hydrometallurgy* **2015**, *157*, 348–355. [CrossRef]
12. Wang, R.C.; Zhai, Y.C.; Xiao-Wei, W.U.; Ning, Z.Q.; Pei-Hua, M.A. Extraction of alumina from fly ash by ammonium hydrogen sulfate roasting technology. *Trans. Nonferr. Met. Soc.* **2014**, *24*, 1596–1603. [CrossRef]
13. Sun, Z.; Li, H.; Bao, W.; Wang, C.; Sun, Z.; Li, H.; Bao, W.; Wang, C. Mineral phase transition of desilicated high alumina fly ash with alumina extraction in mixed alkali solution. *Int. J. Miner. Process.* **2016**, *153*, 109–117. [CrossRef]

14. Gong, B.; Chong, T.; Zhuo, X.; Zhao, Y.; Zhang, J. Mineral changes and trace element releases during extraction of alumina from high aluminum fly ash in Inner Mongolia, China. *Int. J. Coal. Geol.* **2016**, *166*, 96–107. [CrossRef]
15. Yuan, H.; Huang, F.; Wang, L.; Li, H. Study of sinter technique with low alumina-silica ratio. *Guizhou Chem. Ind.* **2004**, *29*, 21–27. (In Chinese)
16. Meher, S.N.; Rout, A.K.; Padhi, B.K. Extraction of alumina from red mud by divalent alkaline earth metal soda ash sinter process. In *Light Metals*; Springer: Cham, Switzerland, 2011; pp. 231–236.
17. Lin, I.J.; Malts, N.; Shindler, Y. The complex chemical treatment of alumina–silica-containing materials. *J. Mater. Synth. Process.* **1998**, *6*, 27–35. [CrossRef]
18. Inada, M.; Eguchi, Y.; Enomoto, N.; Hojo, J. Synthesis of zeolite from coal fly ashes with different silica–alumina composition. *Fuel* **2005**, *84*, 299–304. [CrossRef]
19. Mouhtaris, T.; Charistos, D.; Kantiranis, N.; Filippidis, A.; Kassoli-Fournaraki, A.; Tsirambidis, A. GIS-type zeolite synthesis from Greek lignite sulphocalcic fly ashes promoted by NaOH solutions. *Microporous Mesoporous Mater.* **2003**, *61*, 57–67. [CrossRef]
20. Lin, C.F.; Hsi, H.C. Resource recovery of waste fly ash: Synthesis of zeolite-like materials. *Environ. Sci. Technol.* **1995**, *29*, 1109–1117. [CrossRef] [PubMed]
21. Murayama, N.; Yamamoto, H.; Shibata, J. Mechanism of zeolite synthesis from coal fly ash by alkali hydrothermal reaction. *Int. J. Miner. Process.* **2002**, *64*, 1–17. [CrossRef]
22. Zhang, J.B.; Li, S.P.; Li, H.Q.; He, M.M. Acid activation for pre-desilicated high-alumina fly ash. *Fuel Process. Technol.* **2016**, *151*, 64–71. [CrossRef]
23. Zhu, G.; Wei, T.; Sun, J.; Gong, Y.; Liu, L. Effects and mechanism research of the desilication pretreatment for high-aluminum fly ash. *Energy Fuels* **2013**, *27*, 6948–6954. [CrossRef]
24. Zhang, J.; Li, S.; Li, H.; Wu, Q.; Xi, X.; Li, Z. Preparation of al–si composite from high-alumina coal fly ash by mechanical–chemical synergistic activation. *Ceran. Int.* **2017**, *43*, 6532–6541. [CrossRef]
25. Zhang, J.; Li, H.; Li, S.; Hu, P.; Wu, W.; Wu, Q. Mechanism of mechanical–chemical synergistic activation for preparation of mullite ceramics from high-alumina coal fly ash. *Ceran. Int.* **2018**, *44*, 3884–3892. [CrossRef]
26. He, S.; Li, H.; Li, S.; Li, Y.; Xie, Q. Kinetics of desilication process of fly ash with high aluminum from pulverized coal fired boiler in alkali solution. *Chin. J. Nonferr. Met.* **2014**, *24*, 1888–1894. (In Chinese)

Article

Oxidation of Thiosulfate with Oxygen Using Copper (II) as a Catalyst

Juan Manuel González Lara [1], Francisco Patiño Cardona [2], Antonio Roca Vallmajor [3,*] and Montserrat Cruells Cadevall [3]

1 Departamento de Biotecnología, Universidad Politécnica de Pachuca, Carr. Pachuca-Ciudad Sagahún km 20, Ex-Hacienda de Santa Bárbara, Zampoala Hidalgo 43830, Mexico; joanmanelgl@yahoo.es
2 Ingeniería en Energía, Universidad Metropolitana de Hidalgo, Boulevard acceso a Tolcayuca 1009, Ex-Hacienda San Javier 43860, Mexico; franciscopatinocardona@gmail.com
3 Departament de Ciència de Materials i Química Física, Facultat de Química, Universitat de Barcelona, Martí i Franquès 1, 08028 Barcelona, Spain; mcruells@ub.edu
* Correspondence: roca@ub.edu; Tel.: +34-934021295

Received: 2 March 2019; Accepted: 25 March 2019; Published: 28 March 2019

Abstract: Thiosulfate effluents are generated in the photography and radiography industrial sectors, and in a plant in which thiosulfates are used to recover the gold and silver contained in ores. Similar effluents also containing thiosulfate are those generated from the petrochemical, pharmaceutical and pigment sectors. In the future, the amounts of these effluents may increase, particularly if the cyanides used in the extraction of gold and silver from ores are substituted by thiosulfates, or if the same happens to electronic scrap or in metallic coating processes. This paper reports a study of the oxidation of thiosulfate, with oxygen using copper (II) as a catalyst, at a pH between 4 and 5. The basic idea is to avoid the formation of tetrathionate and polythionate, transforming the thiosulfate into sulfate. The nature of the reaction and a kinetic study of thiosulfate transformation, by reaction with oxygen and Cu^{2+} at a ppm level, are determined and reported. The best conditions were obtained at 60 °C, pH 5, with an initial concentration of copper of 53 ppm and an oxygen pressure of 1 atm. Under these conditions, the thiosulfate concentration was reduced from 1 $g{\cdot}L^{-1}$ to less than 20 ppm in less than three hours.

Keywords: thiosulfate oxidation; kinetics; catalysis; oxygen

---

1. Introduction

In metallurgy, there is growing interest in technology for the extraction of gold and silver that does not use cyanide, for application in minerals or in residues in which it is difficult to recover metals from leaching liquids, and also for environmental purposes. The dissolution of gold and silver by thiosulfate occurs via the formation of the corresponding metal complexes. Technological proposals for implementing this process include the use of a copper (II) salt as an oxidizing agent and ammonia for copper stabilization within the system. One advantage in the use of thiosulfates is the reduction of interaction with other cations, such as copper, arsenic, and antimony, in contrast to the case of cyanidation. The thiosulfate route for the recovery of gold and silver from ores using Cu (II) and ammonia is complex. The leaching rates are acceptable, but the consumption of reagents can be high due to thiosulfate degradation. There are also difficulties encountered during the recovery of the dissolved metals [1–11]. Several authors have proposed modifications of the process reported in the literature. An alternative process includes ferric-ethylenediaminetetraacetic acid (EDTA), ferric-oxalate, and ferric-citrate to avoid the difficulties with the cupric-tetra-amine additive, including excessive thiosulfate oxidation and high reagent costs [12], whereas other authors presented an analysis of the effect of EDTA, thiosulfate, and cupric ions on silver leaching kinetics [13]. At present, however, thiosulfate leaching of minerals is only applied commercially in the Goldstrike deposit of the Barrick Gold Corporation [8].

Meanwhile, every year, millions of tons of waste are generated worldwide from electrical and electronic products. These devices are an important source of secondary raw materials in the form of gold and other metals with high economic values [14,15]. The amount of gold on the printed circuit boards (PCBs) of mobile phones can reach 300–350 g·ton^{-1} [16]. A leaching process using thiosulfate in an ammoniacal medium to recover the gold contained in these PCBs has also been reported [17]. The authors indicated that the leaching system offers promising opportunities for industrial application; moreover, optimization of the leaching system for recovering gold from such PCBs was also carried out [14].

One of the most important uses of thiosulfate solutions has been for application in the photographic industry (in fixing baths and others) to dissolve the silver halide not reduced to metallic silver during photographic or radiographic processes. Today, the industry still generates significant amounts of thiosulfate-based effluents that need to be detoxified for environmental purposes, and to recover the silver [18–21].

In recent decades, for gold and silver coatings, the sulfite-thiosulfate system has been proposed as a non-cyanide coating technology [22–27].

As well as the thiosulfate effluents generated in the photography and radiography industries, the plants that use thiosulfates to recover the gold and silver contained in ores, and also the petrochemical, pharmaceutical, and pigment industries, among others, all produce thiosulfate effluents. In the future, the amounts of these effluents may increase considerably, particularly if cyanides are substituted by thiosulfates in the extraction of gold and silver from their ores, in the recovery of metals contained in electronic scrap, or in metal coating processes. Therefore, an adequate process is needed to degrade the thiosulfate contained in the corresponding effluents.

Different processes for this degradation of thiosulfates have been proposed. A study of the oxidation of thiosulfates by oxygen, using synthetic sphalerite doped with transition metals [28], the oxidation of thiosulfate with H_2O_2 in a catalyzed reaction, being the oxidation products sulfite and sulfate [29], and the oxidation of thiosulfates contained in wastewater, with oxygen in the presence of UV light [30], have been carried out.

Our research team studied the transformation of thiosulfate using copper sulfate solutions. The amount of copper (II) salt used was almost the stoichiometric quantity for the formation of 1 mol CuS and 1 mol sulfates per mol thiosulfate. The transformation of thiosulfate using copper (II) sulfate was applied to an industrial fixing bath from the photographic industry; the final effluent contained less than 10 mg L^{-1} of thiosulfate [31].

This paper reports a study of the oxidation of thiosulfate to sulfate, with oxygen using copper (II) as a catalyst. The reaction was developed at a pH between 4 and 5 and the idea is to avoid the formation of tetrationate or polythionate. The nature of the reaction and its kinetics, with oxygen and in this case with Cu (II) at the ppm level, are determined. The effects of temperature and the initial concentrations of H_3O^+, Cu^{2+}, and $S_2O_3{}^{2-}$ on the thiosulfate transformation are also determined; and the effect of the partial pressure of oxygen is studied.

2. Materials and Methods

2.1. Materials

In the study of thiosulfate degradation, sodium thiosulfate solutions prepared from this salt, with 99% purity, and pentahydrate copper sulfate of the same purity were used. Oxygen with a purity of ≥99.995% (mol/mol) was also used.

2.2. Experimental Procedure for Thiosulfate Transformation Using Copper Sulfate

The experiments were carried out in a 0.5 L flat-bottom temperature-controlled reactor with magnetic stirring. The pH was kept constant by adding the necessary amounts of 0.5 mol·L^{-1} H_2SO_4 solution and 0.5 mol·L^{-1} NaOH solution.

In each experiment, the stirring rate, pH, air/oxygen flow, and temperature were adjusted to the values required in each case. Samples were taken at different times and filtered; and the remaining thiosulfate in the liquids was analyzed by iodometry (iodine 0.1 mol·L^{-1} was used for this purpose and a starch solution was added to determine the end point). Some samples were also analyzed via ionic chromatography. During the process, the solids generated were characterized by X-ray diffraction (XRD, PANanalytical, Almelo, The Netherlands) as well as by scanning electron microscopy (SEM, JEOL, Tokyo, Japan), and energy-dispersive X-ray analysis (EDS, Oxford Instruments, Abingdon, England). The rates of these processes are indicated from the k_{exp} values obtained in each experiment (slope of the graph: conversion, X, versus time); the conversion was defined according to the expression:

$$X = [S_2O_3{}^{2-}]_{transformed}/[S_2O_3{}^{2-}]_{initial.} \quad (1)$$

3. Results and Discussion

Preliminary experiments involving thiosulfate oxidation by oxygen, but without the presence of Cu (II), led to very slow rates for the process.

3.1. The Nature of the Thiosulfate Transformation by Reaction with Cu (II) Sulfate and Oxygen

Two experiments involving thiosulfate degradation using Cu (II) were carried out under the following conditions: temperature, 60 °C; stirring rate, 500 min^{-1}; initial concentration of thiosulfate, 1 g·L^{-1}; initial concentration of Cu^{2+}, 8.4×10^{-4} mol·L^{-1} (0.053 g·L^{-1}); and oxygen pressure, 1 atm. The first experiment was carried out at pH 5, and the second at pH 4.

Figure 1 shows the results for thiosulfate transformation at the indicated pH values. The thiosulfate degradation was almost total for a reaction time of 160 min. Meanwhile, two different rates were obtained: a slower one during the first part of the process, and a faster one during the second part. The values obtained were similar in both cases, pH 5 and pH 4. At a conversion of nearly to 0.40 (both experiments), the change in the slope was observed as the reaction rate increased, and a black solid was formed: CuS as confirmed by XRD (Figure 2). Therefore, it seems that the presence of this CuS precipitate enhanced the reaction rate, indicating a catalytic effect on thiosulfate degradation of the contact with copper sulfide surface. The precipitate of copper sulfide consisted of aggregates of crystals with 1 μm size (SEM-EDS).

Final liquids were analyzed by ion chromatography, indicating that thiosulfates were oxidized to sulfates under the experimental conditions employed (see Figure 3). In the pH interval between 4 and 5, sulfates were detected; any other species, such as tetrathionate or polytionate, were not detected.

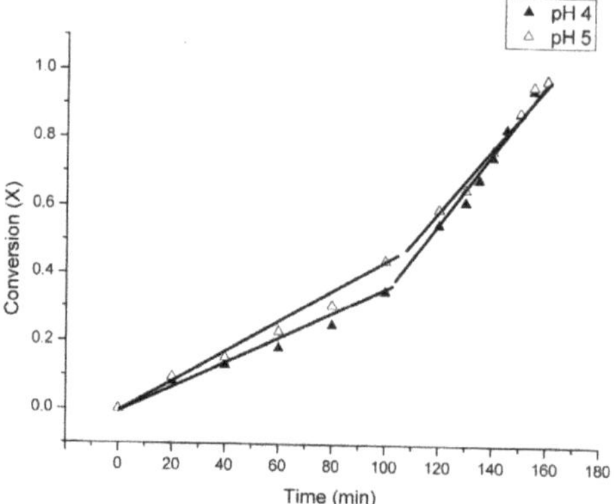

Figure 1. Nature of the reaction for thiosulfate transformation using Cu^{2+} and oxygen: $[S_2O_3^{2-}]_{initial}$, 1 g·L^{-1}; $[Cu^{2+}]_{initial}$, 0.053 g·L^{-1}; $pH_{constant}$, 4 or 5; 60 °C.

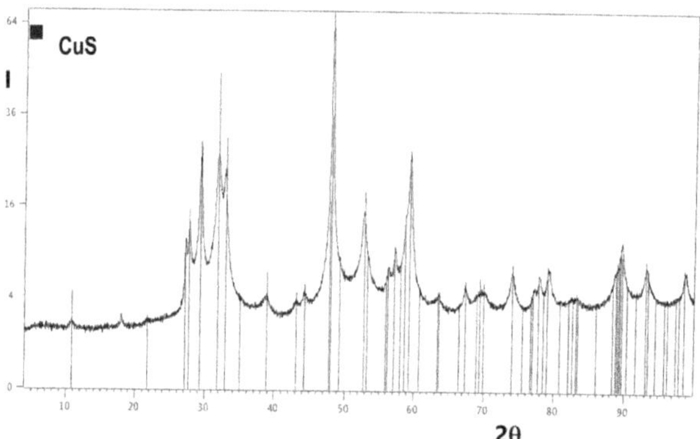

Figure 2. Nature of the reaction: X-ray diffractogram of solid obtained during thiosulfate transformation: identified as CuS.

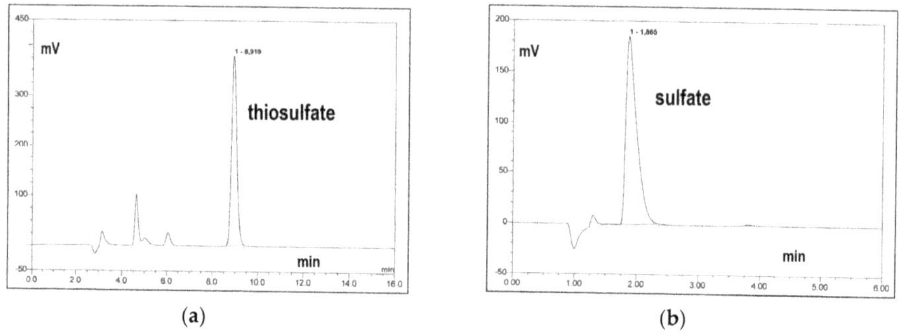

Figure 3. Chromatograms: (**a**) Initial solution; (**b**) Final solution.

When CuS had formed (second step), hydronium ions were generated and pH was kept constant by the addition of sodium hydroxide solution.

The amount of thiosulfate remaining in the solution at the end of the experiment was less than 20 mg·L^{-1} in both cases. Table 1 includes the slopes of the graphs $X_{thiosulfate}$ vs. time (k_{exp}), which is a measure of the transformation rate.

Table 1. Values of k_{exp} of thiosulfate transformation in two steps.

pH	Time (min) First Step	Time (min) Second Step	k_{exp} (min^{-1}) First Step	k_{exp} (min^{-1}) Second Step
5	0–80	80–160	0.0040	0.0070
4	0–100	100–160	0.0033	0.0110

Iodometry established the absence of sulfites, while tetrathionates or polythionates were not detected by chromatography. Thus, the initial thiosulfates were transformed to sulfates.

Under the employed experimental conditions, the reaction produced in the zone where the change in the slope occurs could be as follows:

$$S_2O_3{}^{2-}{}_{(aq)} + Cu^{2+}{}_{(aq)} + 3H_2O \Rightarrow SO_4{}^{2-}{}_{(aq)} + CuS_{(s)} + 2H_3O^+{}_{(aq)}. \quad (2)$$

This process is quite similar to the degradation process using large amounts of copper ions (0.21–0.85 $g{\cdot}L^{-1}$ Cu). In that case, sulfates and copper sulfide were formed in a similar amount in mol, when >0.57 $g{\cdot}L^{-1}$ Cu were used. After this, the copper sulfide formed, and had to be oxidized to copper sulfate prior to being reused in the next cycle [31]. By using 0.053 $g{\cdot}L^{-1}$ Cu (in the present work) and oxygen, the reaction rates are slower, but the transformation of thiosulfate to sulfate takes place in more than 90% yield. In this case, the amount of copper sulfide that must be oxidized to sulfate is much lower than the work in which large amounts of copper were used.

Another difference between these two procedures is that when we used very low amounts of copper salts and oxygen, two rates were detected; whereas, when using larger amounts of copper (without oxygen), only one rate value (k_{exp} = 0.013 min^{-1}) was determined, and the thiosulfate degradation was completed in a shorter time. The work cited [31] also demonstrated that, in experiments carried out at pH 6 and pH 10, no thiosulfate transformation to sulfate was detected. Meanwhile, thiosulfates were easily decomposed at pH $\leq$ 4, giving $HSO_3{}^-$, SO_2, and elemental sulfur. For these reasons, the kinetic study in this work was carried out in the pH interval between 4 and 5.

3.2. Kinetic Study of the Transformation of Thiosulfate by Reaction with Oxygen and Cu^{2+}

The effects of temperature (40–80 °C), concentration of H_3O^+ (1.0×10^{-4}–1.0×10^{-5} $mol{\cdot}L^{-1}$), initial Cu^{2+} (4.2×10^{-4}–1.68×10^{-3} $mol{\cdot}L^{-1}$), and initial $S_2O_3{}^{2-}$ (4.5×10^{-3}–9×10^{-3} $mol{\cdot}L^{-1}$), as well as an oxygen pressure of 0.2–1 atm, were determined.

Figure 4 shows the plot of conversion vs. time at different temperatures. Table 2 includes the experimental values of k_{exp} for these experiments; the table includes the corresponding values corrected by the oxygen concentration at each temperature, because the oxygen concentration varies with temperature [32]. The activation energy was calculated according to the following expression:

$$(k_{exp}/[O_2]) = A\exp(E_a/RT). \quad (3)$$

Figure 5 includes the plot of $\ln(k_{exp}/[O_2])$ vs. $1000/T$. Two apparent activation energies E_a were obtained: for the first step, 41 $kJ{\cdot}mol^{-1}$ and for the second, 33 $kJ{\cdot}mol^{-1}$.

For concentrations of H_3O^+ varying from 1.0×10^{-4} to 1.0×10^{-5} $mol{\cdot}L^{-1}$ H_3O^+ (pH between 4 and 5, Figure 6), the reaction rate, k_{exp}, in the first step was between 0.0033 and 0.040 min^{-1}; and, in the second step, it was between 0.070 and 0.0110. All the values obtained are similar; consequently, an apparent reaction order of $\approx$0 with respect to the hydronium concentration was obtained.

The effect of the copper concentration was studied in the interval between 1.68×10^{-3} and 4.20×10^{-4} $mol{\cdot}L^{-1}$ Cu^{2+}, that is, between 0.11 and 0.027 $g{\cdot}L^{-1}$ Cu^{2+}; Figure 7 shows the results. Figure 8 is a graph corresponding to the experiment performed at 1.68×10^{-3} M Cu^{2+}, to show the appearance of two stages, as in the other experiments; this does not appear in Figure 7. Table 3 includes the experimental constants for these experiments.

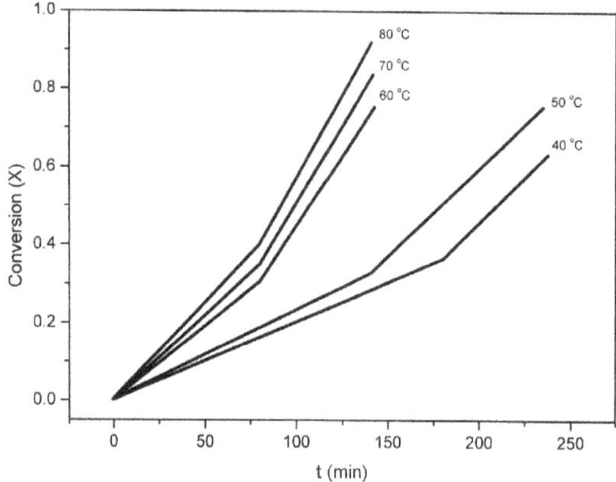

Figure 4. Effect of temperature on thiosulfate transformation: $[S_2O_3^{2-}]_{initial}$, 1 g·L^{-1}; $[Cu^{2+}]_{initial}$, 0.053 g·L^{-1}; $pH_{constant}$, 5; O_2 pressure, 1 atm.

Table 2. Effect of temperature on thiosulfate transformation.

Temp. (°C)	1000/*T* (K^{-1})	$k_{exp(1)}$ (min^{-1})	$k_{exp(1)}/[O_2]$ (min^{-1}/mol·m^{-3})	$Ln(k_{exp(1)}/[O_2])$	$k_{exp(2)}$ (min^{-1})	$k_{exp(2)}/[O_2]$ (min^{-1}/mol·m^{-3})	$Ln(k_{exp(2)}/[O_2])$
40	3.195	0.0020	2.08×10^{-3}	−6.177	0.0045	4.67×10^{-3}	−5.366
50	3.096	0.0022	2.65×10^{-3}	−5.933	0.0057	6.87×10^{-3}	−4.981
60	3.003	0.0038	5.35×10^{-3}	−5.231	0.0075	0.0106	−4.550
70	2.915	0.0042	7.24×10^{-3}	−4.928	0.0080	0.0138	−4.280
80	2.833	0.0050	0.0116	−4.457	0.0086	0.020	−3.912

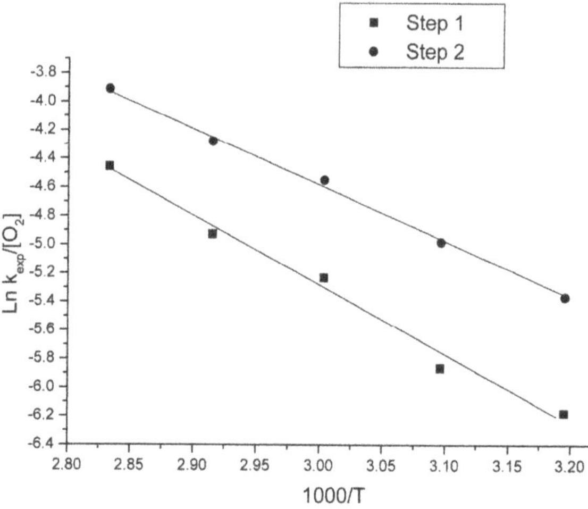

Figure 5. Apparent activation energy (E_a) for thiosulfate transformation (first step): 41 kJ·mol^{-1}, and (second step): 33 kJ·mol^{-1}.

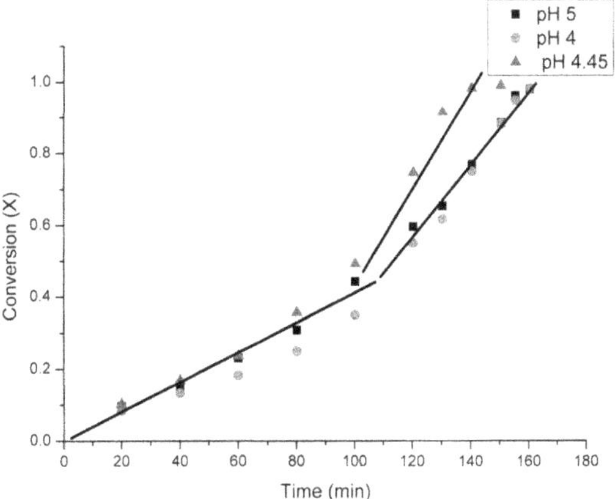

Figure 6. Effect of H_3O^+ concentration on thiosulfate transformation: $[S_2O_3{}^{2-}]_{\text{initial}}$, 1 g·L^{-1}; $[Cu^{2+}]_{\text{initial}}$, 0.053 g·L^{-1}; 60 °C; O_2 pressure, 1 atm.

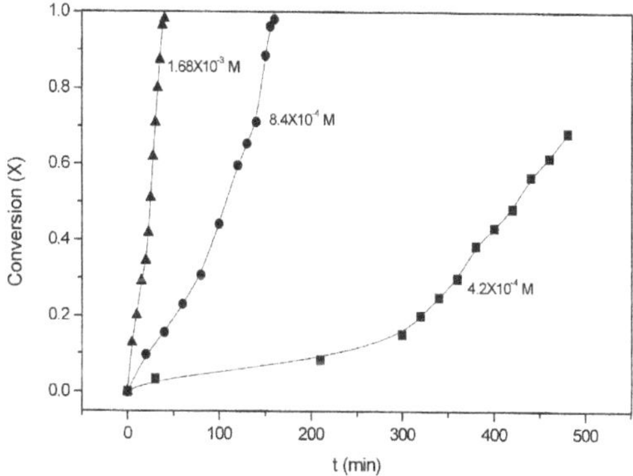

Figure 7. Effect of initial Cu^{2+} concentration on thiosulfate transformation: $[S_2O_3{}^{2-}]_{\text{initial}}$, 1 g·L^{-1}; pH_{constant}, 5; 60 °C; O_2 pressure, 1 atm.

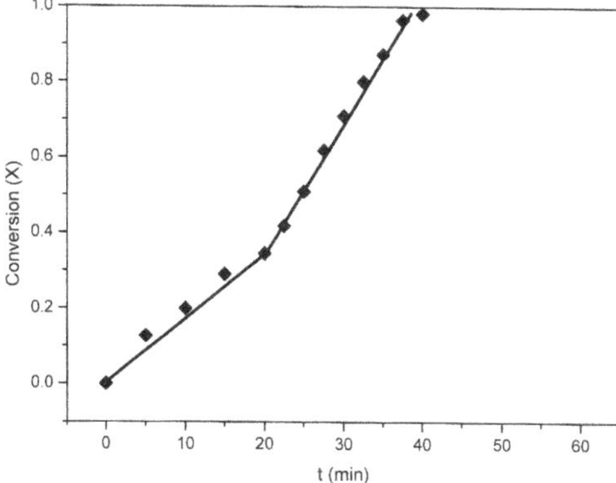

Figure 8. Effect of initial Cu^{2+} concentration on thiosulfate transformation: Experiment performed at 1.68×10^{-3} mol·L^{-1} Cu^{2+} (0.107 g·L^{-1}).

Table 3. Effect of copper concentration: values of the experimental constants.

$[Cu^{2+}]$ ($mol{\cdot}L^{-1}$)	$k_{exp(1)}$ (min^{-1})	$k_{exp(2)}$ (min^{-1})
4.2×10^{-4}	0.00044	0.0030
8.4×10^{-4}	0.0030	0.011
1.68×10^{-3}	0.017	0.034

Figure 9 is a graph of log k_{exp} versus log initial $[Cu^{2+}]$. Apparent reaction orders of 2.5 for the first step and 1.75 for the second step were obtained. These values show the important effect that the initial concentration of copper (II) has on the rate of transformation of thiosulfates in the interval of copper concentrations from 4.2×10^{-4} to 1.68×10^{-3} $mol{\cdot}L^{-1}$.

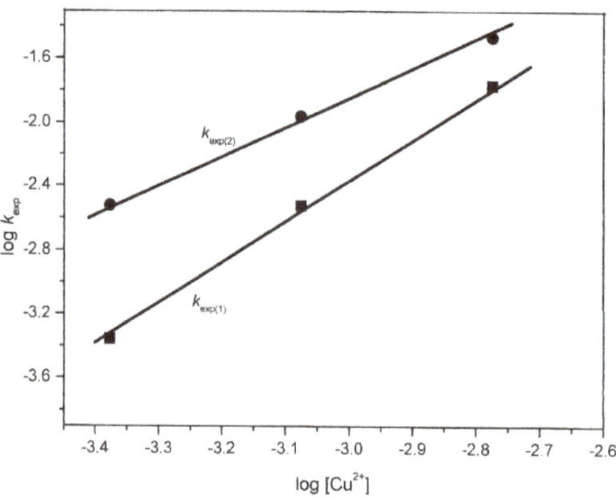

Figure 9. Effect of the initial Cu^{2+} concentration on thiosulfate transformation: order of the reaction.

The effect of the initial concentration of thiosulfates on their transformation rate was determined for the interval 4.5×10^{-3}–9×10^{-3} $mol{\cdot}L^{-1}$ (0.50–1 $g{\cdot}L^{-1}$); Figure 10 includes the results. As the reaction rate decreases, as the initial thiosulfate concentration increases. When the molar ratio $S_2O_3^{2-}/Cu^{2+}$ is ≤ 8, only one k_{exp} value was obtained; for values of the molar ratio >10, two slopes appeared. Table 3 includes values of the log of the initial thiosulfate concentration and the log of the different values of k_{exp}.

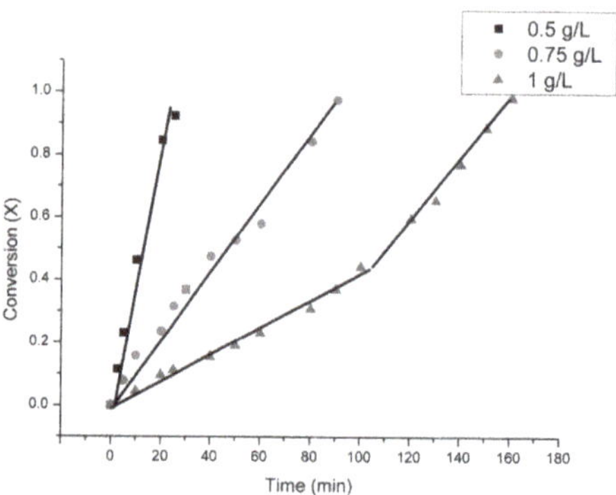

Figure 10. Effect of initial $S_2O_3^{2-}$ concentration on thiosulfate transformation: $[Cu^{2+}]_{initial}$, 0.053 $g{\cdot}L^{-1}$; $pH_{constant}$, 5; 60 °C; O_2 pressure, 1 atm.

Table 4, lines 1 and 2, contain the k_{exp} values corresponding to the experiments in which 0.5 and 0.75 g·L^{-1} of thiosulfate were used; line 3 includes the average value of k_{exp} obtained in the 1 g·L^{-1} thiosulfate experiment (under these conditions, a change of slope was detected). Lines 4 and 5 correspond to the k_{exp} values of the first and second stage, respectively, of the experiment with 1 g·L^{-1} of thiosulfates. The apparent reaction order, considering the first three lines, is −2.8; if the experiment at 1 g·L^{-1} is considered in two steps, the apparent order of the first step is −3.4 and the apparent order of the second is −2.6. When the thiosulfate concentration was increased, the reaction rate decreased significantly, because of the increases in the thiosulfate/Cu(II) ratio.

Table 4. Reaction order with respect to the initial thiosulfate concentration.

Line	$[S_2O_3^{2-}]$ (mol·L^{-1})	log $[S_2O_3^{2-}]$ (mol·L^{-1})	k_{exp} (min^{-1})	log k_{exp} (min^{-1})
1	4.46×10^{-3}	−2.351	0.0410	−1.387
2	6.70×10^{-3}	−2.174	0.0098	−2.009
3	8.92×10^{-3}	−2.050	0.0060	−2.222
4	8.92×10^{-3}	−2.050	0.0040	−2.398
5	8.92×10^{-3}	−2.050	0.0070	−2.155

An additional experiment was carried out by decreasing the $S_2O_3^{2-}/Cu^{2+}$ ratio, using 3.6×10^{-2} mol·L^{-1} thiosulfate (4 g·L^{-1} thiosulfate) and an initial Cu^{2+} concentration of 2.5×10^{-3} mol·L^{-1} (initial thiosulfate/copper molar ratio of 14.2). The behavior was quite similar to that shown in Figure 8 (initial copper concentration 4.2×10^{-4} mol·L^{-1}). The rate values obtained were: from 0 to 270 min, k_{exp1} = 0.0020 and k_{exp2} = 0.0083.

The effect of the partial pressure of oxygen was determined in the interval between 0 and 1 atm pressure; Figure 11 shows the results. The reaction rate increases as the partial pressure of oxygen increases. In the experiment without oxygen, a conversion of 0.33 was achieved after 480 min reaction time (First step: k_{exp} = 0.0006 min^{-1}). When air was used (partial pressure of oxygen = 0.2), a conversion of 0.50 was achieved at the same time (First step: k_{exp} = 0.0010 min^{-1}). Finally, with oxygen at a partial pressure of 1 atm, a conversion of 0.98 was obtained in 160 min (k_{exp} = 0.0040 min^{-1}, first step; and k_{exp} = 0.0070 min^{-1}, second step). Only in the last experiment was the degradation reaction almost completed. Consequently, an average value of reaction order near to 1 was obtained with respect to the partial pressure of oxygen. The reaction rate was multiplied by 4 when oxygen was used instead of air, i.e., approximately the same ratio of the oxygen partial pressure in air or in pure oxygen.

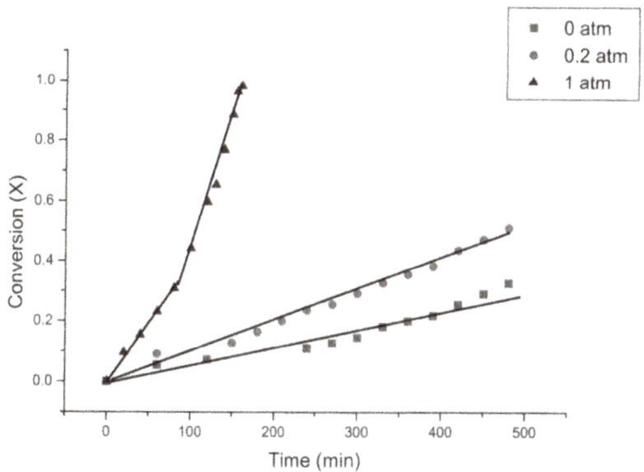

Figure 11. Effect of the partial pressure of oxygen on thiosulfate transformation: $[S_2O_3^{2-}]$, 1 g·L^{-1}; $[Cu^{2+}]_{initial}$, 0.053 g·L^{-1}; pH$_{constant}$, 5; 60 °C.

The transformation of thiosulfate to sulfate using oxygen and copper (II) in dilute solution (25–100 ppm), at pH 5 and 60 °C, has potential applications in treating effluents for thiosulfate degradation. The small amounts of copper sulfide generated can be oxidized to copper sulfate and recycled to the next cycle of thiosulfate degradation.

4. Conclusions

(1) In the degradation of thiosulfate by oxygen and small amounts of Cu^{2+} at pH between 4 and 5, a change in the slope of the graph of reaction rate (k_{exp}) versus time was observed. This coincided with a precipitate of CuS being formed, and the reaction rate increased from this point. This implies a catalytic effect of contact with CuS on the thiosulfate degradation.

(2) The reaction produced in the zone where the change in the slope occurs could be as follows:

$$S_2O_3{}^{2-}{}_{(aq)} + Cu^{2+}{}_{(aq)} + 3H_2O \Rightarrow SO_4{}^{2-}{}_{(aq)} + CuS_{(s)} + 2H_3O^+{}_{(aq)}.$$

Analysis of the final liquids carried out by chromatography confirmed that the thiosulfates were oxidized to sulfates under the employed experimental conditions. The amount of thiosulfate remaining in solution at the end of the process was less than 20 $mg \cdot L^{-1}$.

(3) In the kinetic study of thiosulfate degradation, two apparent activation energies were obtained: 41 $kJ \cdot mol^{-1}$ (in the first step, until a conversion of nearly 0.35) and 33 $kJ \cdot mol^{-1}$ (in the second step after CuS formation). The effect of temperature on the degradation process using oxygen and Cu (II) at the ppm level is less than that observed when using larger amounts of Cu (II) without oxygen (98 $kJ \cdot mol^{-1}$).

(4) For a concentration of Cu (II) varying between 1.68×10^{-3} and 4.20×10^{-4} $mol \cdot L^{-1}$ Cu^{2+} (107–27 ppm Cu in solution), the reaction rate increased with the Cu (II) concentration, giving an apparent reaction order near to 2 (2.5 for the first step and 1.75 for the second step).

(5) For a concentration of thiosulfate varying between 4.5×10^{-3} and 9×10^{-3} $mol \cdot L^{-1}$ $S_2O_3{}^{2-}$, only one reaction rate (slope) was detected for molar ratios thiosulfate/Cu $\leq$8. Two rates (slopes) were detected for values of this ratio higher than 10. Apparent reaction orders of −3.4 (first step) and −2.6 (second step) were obtained, indicating that, when the thiosulfate concentration increases, the reaction rate decreases significantly, because of the increases in the thiosulfate/Cu(II) ratio.

(6) The effect of the partial pressure of oxygen was determined in the interval between 0 and 1 atm pressure. The reaction rate increases as the partial pressure of oxygen increases. An average apparent reaction order near to 1 was obtained.

(7) The transformation of thiosulfate to sulfate by oxygen and a dilute copper (II) solution (25–100 ppm), at pH 5 and 60 °C, has potential applications in treating effluents for thiosulfate degradation. The CuS generated during the process can be oxidized and recycled to the next cycle of thiosulfate degradation.

Author Contributions: A.R.V., M.C.C. and F.P.C. conceived and designed the experiments; J.M.G.L. performed the experiments; A.R.V., M.C.C., J.M.G.L. and F.P.C. analyzed the data; and A.R.V. and M.C.C. wrote the paper.

Acknowledgments: The authors wish to thank the *Centres Científics i Tecnològics* of the Universitat de Barcelona for their assistance with this work.

Conflicts of Interest: The authors declare no conflict of interest.

List of Symbols

Conversion	X (dimensionless)
k_{exp}	reaction rate (min^{-1})
Activation energy	Ea ($kJ \cdot mol^{-1}$)

References

1. Breuer, P.L.; Jeffrey, M.I. Thiosulfate leaching kinetics of gold in the presence of copper and ammonia. *Min. Eng.* **2000**, *13*, 1071–1081. [CrossRef]
2. Li, J.; Miller, J.D.; le Vier, M.; Wan, R.Y. The ammoniacal thiosulfate system for precious metal recovery. In Proceedings of the XIX International Mineral Processing Congress, Precious Metals Processing and Mineral Waste and the Environment, San Francisco, CA, USA, 22–27 October 1995; pp. 37–42, ISBN 0-87335-142-8.
3. Senanayake, G. Analysis of reaction kinetics, speciation and mechanism of gold leaching and thiosulfate oxidation by ammoniacal copper (II) solutions. *Hydrometallurgy* **2004**, *75*, 55–75. [CrossRef]
4. Zhang, S.; Nicol, J.M. An electrochemical study of the dissolution of gold in thiosulfate solutions. Part I. Alkaline solutions. *J. Appl. Electrochem.* **2003**, *33*, 767–775.
5. Zhang, S.; Nicol, J.M. An electrochemical study of the dissolution of gold in thiosulfate solutions. Part II. Effect of copper. *J. Appl. Electrochem.* **2005**, *35*, 339–345. [CrossRef]
6. Syed, S. Recovery of gold from secondary sources-A review. *Hydrometallurgy* **2012**, *115–116*, 30–51. [CrossRef]
7. Zhang, X.M.; Senanayake, G. A review of ammoniacal thiosulfate leaching of gold: An update useful for further research in Non-cyanide gold lixiviants. *Miner. Process Extract. Metall. Rev.* **2016**, *37*, 385–411. [CrossRef]
8. Xu, B.; Kong, W.; Li, Q.; Yang, Y.; Jiang, T.; Liu, X. A Review of Thiosulfate Leaching of Gold: Focus on Thiosulfate Consumption and Gold Recovery from Pregnant Solution. *Metals* **2017**, *7*, 222. [CrossRef]
9. Celep, O.; Altinkaya, P.; Yazici, E.Y.; Deveci, H. Thiosulphate leaching of silver from an arsenical refractory ore. *Min. Eng.* **2018**, *122*, 285–295. [CrossRef]
10. Juárez, J.C.; Patiño, F.; Roca, A.; Teja, A.M.; Reyes, I.A.; Reyes, M.; Pérez, M.; Flores, M.U. Determination of dissolution rates of Ag contained in metallurgical and mining residues in the S_2O_3-O_2-Cu^{2+}system: Kinetic analysis. *Minerals* **2018**, *8*, 309. [CrossRef]
11. Sitando, O.; Senanayake, G.; Dai, X.; Nikoloski, A.N.; Breuer, P. A review of factors affecting gold leaching in non-ammoniacal thiosulfate solutions including degradation and in-situ generation of thiosulfate. *Hydrometallurgy* **2018**, *78*, 151–175. [CrossRef]
12. Deutsch, J.L.; Dreisinger, D.B. Silver sulfide leaching with thiosulfate inthe presence of additives. Part I: Ferric complexes and the application to silver sulfide ores. *Hydrometallurgy* **2013**, *137*, 165–172. [CrossRef]
13. Puente-Siller, D.M.; Fuentes-Aceituno, J.C.; Nava-Alonso, F. A kinetic-thermodynamic study of silver leaching in thiosulfate-copper-ammonia-EDTA solutions. *Hydrometallurgy* **2013**, *134–135*, 124–131. [CrossRef]
14. Ha, V.H.; Lee, J.; Huynh, T.H.; Jeong, J.; Pandey, B.D. Optimizing the thiosulfate leaching of gold from printed circuit boards of discarded mobile phone. *Hydrometallurgy* **2014**, *149*, 118–129. [CrossRef]
15. Camelino, S.; Rao, J.; Padilla, R.L.; Lucci, R. Initial studies about gold leching from printed circuit boardss (PCB'S) of waste cell phones. *Procedia Mater. Sci.* **2015**, *9*, 105–112. [CrossRef]
16. Hagelüken, C.; Corti, C.W. Recycling of gold from electronics: Cost effective use through 'Design for Recycling'. *Gold Bull.* **2010**, *43*, 209–220. [CrossRef]
17. Kasper, A.C.; Veit, H.M. Leaching of Gold from Printed Circuit Boards Scrap of Mobile Phones. In *Energy Technology*; Springer: Cham, Switzerland, 2015; ISBN 978-3-319-48602-4.
18. Bas, A.D.; Yazici, E.Y.; Deveci, H. Recovery of silver from X-ray film processing effluents by hydrogen peroxide treatment. *Hydrometallurgy* **2012**, *121–124*, 22–27. [CrossRef]
19. Mahdizadeh, F.; Eskandarian, M.; Zabarjadi, J.; Ehsani, A.; Afshar, A. Silver recovery from radiographic film processing effluents by hydrogen peroxide: Modeling and optimization using response surface methodology. *Korean J. Chem. Eng.* **2014**, *31*, 74–80. [CrossRef]
20. Prado, P.F.A.; Ruotolo, L.A.M. Silver recovery from simulated photographic baths by electrochemical deposition avoiding Ag_2S formation. *J. Environ. Chem. Eng.* **2016**, *4*, 3283–3292. [CrossRef]
21. Labra, M.P.; Pérez, M.R.; Serrano, J.A.R.; Dávila, D.O.A.; Hernández, F.R.B.; Thangarasu, P. Silver cementation with zinc from residual X ray fixer, experimental and thermochemical study. In *Characterization of Minerals, Metals, and Materials*; Springer: Cham, Switzerland, 2016; pp. 605–613.
22. Blair, A. Silver plating. *Met. Finish.* **1995**, *93*, 290–297. [CrossRef]
23. Osaka, T.; Kodera, A.; Misato, T.; Homma, T.; Okinaka, Y. Electrodeposition of soft gold from thiosulfate-sulfite bath for electronics applications. *J. Electrochem. Soc.* **1997**, *144*, 3462–3469. [CrossRef]

24. Ando, S.; Inoue, T.; Okudaira, H.; Takehara, Y.; Ushio, J.; Ohta, T.; Yamamoto, H. Super stable non-cyanide electroless gold plating bath which has been applied to advanced board manufacture. In Proceedings of the IEMT/IMC Symposium, 1st Joint International Electronic Manufacturing Symposium and the International Microelectronics Conference, Sonic City-Omiya, Tokyo, Japan, 16–18 April 1997; pp. 220–225.
25. Berger, S.; Hoffacker, G. Novel and innovative non-cyanide silver process: A report on commercial plating experiences. *Plat. Surf. Finish.* **2005**, *92*, 46–51.
26. Ren, F.-Z.; Yin, L.-T.; Wang, S.-S.; Volinsky, A.A.; Tian, B.-H. Cyanide-free silver electroplating process in thiosulfate bath and microstructure analysis of Ag coatings. *Trans. Nonferr. Met. Soc. China* **2013**, *23*, 3822–3828. [CrossRef]
27. Sabir, S. *Silver Recovery from Assorted Spent Sources: Toxicology of Silver Ions*; World Scientific Publishing Europe Ltd.: London, UK, 2018; Chapter 1.
28. Xu, Y.; Shoonen, M.A.M. Thiosulfate oxidation: Catalysis of synthetic sphalerite doped with transition metals. *Geochim. Cosmochim. Acta* **1996**, *60*, 4701–4710. [CrossRef]
29. Chatterjee, D.; Shome, S.; Jaiswal, N.; Moi, S.C. Mechanism of the oxidation of thiosulfate with hydrogen peroxide catalyzed by aqua-ethylenediaminetetraacetatoruthenium (III). *J. Mol. Catal. A Chem.* **2014**, *386*, 1–4. [CrossRef]
30. Ahmad, N.; Ahmad, F.; Khan, I.; Khan, A.D. Studies on the Oxidative Removal of Sodium Thiosulfate from Aqueous Solution. *Arab. J. Sci. Eng.* **2014**, *40*, 289–293. [CrossRef]
31. González-Lara, J.M.; Roca, A.; Cruells, M.; Patiño, F. The oxidation of thiosulfates with copper sulfate. Application to an industrial fixing bath. *Hydrometallurgy* **2009**, *95*, 8–14. [CrossRef]
32. Lourié, Y. *Aide-Mémorie de Chimie Analytique*; URSS: Moscou, Russia, 1975; pp. 114–115.

Article

High-Pressure Oxidative Leaching and Iodide Leaching Followed by Selective Precipitation for Recovery of Base and Precious Metals from Waste Printed Circuit Boards Ash

Altansukh Batnasan [1,*], Kazutoshi Haga [1], Hsin-Hsiung Huang [2] and Atsushi Shibayama [1,*]

1 Graduate School of International Resource Sciences, Akita University, 1-1 Tegata-Gakuen machi, Akita 010-8502, Japan; khaga@gipc.akita-u.ac.jp

2 Department of Metallurgy and Materials Engineering, Montana Technological University, 1300 West Park Street, Butte, MT 59701, USA; hhuang@mtech.edu

* Correspondence: altansux@gipc.akita-u.ac.jp or altansukh.b2008@gmail.com (A.B.); sibayama@gipc.akita-u.ac.jp (A.S.); Tel.: +81-18-889-3296 (A.B.); +81-18-889-3051 (A.S.)

Received: 1 March 2019; Accepted: 18 March 2019; Published: 20 March 2019

Abstract: This paper deals with the recovery of gold from waste printed circuit boards (WPCBs) ash by high-pressure oxidative leaching (HPOL) pre-treatment and iodide leaching followed by reduction precipitation. Base metals present in WPCB ash were removed via HPOL using a diluted sulfuric acid solution at elevated temperatures. Effects of potassium iodide concentration, hydrogen peroxide concentration, sulfuric acid concentration, leaching temperature, and leaching time on gold extraction from pure gold chips with $KI–H_2O_2–H_2SO_4$ were investigated. The applicability of the optimized iodide leaching process for the extraction of gold from the leach residue obtained after HPOL were examined at different pulp densities ranging from 50 g/t to 200 g/t. Results show that the removal efficiency was 99% for Cu, 95.7% for Zn, 91% for Ni, 87.3% for Al, 82% for Co, and 70% for Fe under defined conditions. Under the optimal conditions, the percentage of gold extraction from the gold chips and the residue of WPCBs was 99% and 95%, respectively. About 99% of the gold was selectively precipitated from the pregnant leach solution by sequential precipitation with sodium hydroxide and L-ascorbic acid. Finally, more than 93% of gold recovery was achieved from WPCB ash by overall combined processes.

Keywords: waste printed circuit board; gold; iodide; iodine; ascorbic acid; leaching; precipitation

1. Introduction

Hydrometallurgical processing is often used to produce metals from complex ores, concentrates, mine tailings, and secondary sources containing metals, termed as raw materials that are difficult to treat by conventional mineral processing and pyrometallurgical methods [1–3]. The leaching process is an essential step in hydrometallurgical processing and can extract metals from raw materials containing metals using a solvent. A key challenge in the leaching process is to meet selective extraction of a metal of interest from raw material avoiding some issues associated with the further processing of the pregnant leach solution [4,5].

Printed circuit boards (PCBs) are the essential building block in the majority of electric and electronic equipment (EEE) and account for 3–6% of the total constitution of EEE [6,7]. Various kinds of metals involving precious metals (Au, Ag, and Pd), base metals (Cu, Fe, Al, Ni, and Co), and toxic metals (Hg, Pb, Cr and Cd), and different types of plastic materials including non-flame retardant and flame retardant polymers are used in manufacturing of PCBs in different ratios depending on the products [8,9]. It means that waste printed circuit boards (WPCBs) generated from discarded electric and electronic equipment (EEE) can be considered as a secondary source for metals. Because the proportion of valuable metals (Au, Ag, Pd, Pt, Cu, Ni, Co, Zn, and Al) is several times higher than those in their primary ore minerals [8–11], the recovery of valuable metals from WPCBs is the real challenge to maintain their supply chains and reduce the environmental consequences of metal mining. Therefore, over the years a variety of methods such as mechanical, hydrometallurgical, bio-metallurgical, and pyrometallurgical technologies have been used to recover valuable metals from WPCBs [12–14]. Among them, highly selective separation methods, such as hydrometallurgical processes, are more suitable due to the diverse and complex characteristics of the waste [2,15]. The recovery of precious metals from WPCBs has been extensively investigated because of their high economic values compared to other metals [16].

Cyanide leaching has been employed for over a century in gold extraction industries all over the world [17,18]. Nonetheless, in gold hydrometallurgy, a great deal of research has been conducted on the replacement of cyanide with alternative reagents such as thiosulfate, thiourea, and halides due to the environmental and human health risks associated with cyanide toxicity [19–22]. Among them, an iodine–iodide solution is a highly selective and efficient leaching agent for the recovery of gold from gold ores and WPCBs [21,23–25]. Several decades ago, the first patented process by Homick et al. (1976) used an iodine–iodide solution and a reducing agent (hydrazine) to recover gold from PCBs [26]. Subsequently, numerous studies have been devoted to developing the iodine–iodide leaching process for the recovery of gold from gold chips, gold ores, and WPCBs [27–29]. Nevertheless, the main obstacles of the leaching process are high reagent consumptions, higher reagent costs, and the complex chemistry of gold leaching in iodide solution. For the last several decades, few authors have studied the dissolution of gold in iodide solution with several oxidants namely hypochlorite, oxygen, and hydrogen peroxide, respectively to reduce iodine consumption [30–32]. Although several studies have focused on the dissolution of gold in iodine–iodide and iodide solutions with different oxidants, much less attention has paid to recover gold from pregnant leach solution (PLS) resulted from leaching [33–35]. Therefore, extensive studies on the process optimization, the dissolution mechanism of gold in iodide solution, and gold recovery from PLS are necessary.

In this study, gold recovery from the WPCBs ash by a two-step leaching process involving HPOL and iodide leaching followed by reductive precipitation was investigated. The removal of base metals from the WPCBs ash by high-pressure oxidative leaching was examined at different temperatures ranging from 100 °C to 180 °C. The dissolution behavior of gold from gold chips in iodide solutions with hydrogen peroxide and sulfuric acid was studied aiming to optimize the leaching process. Various experimental parameters such as iodide concentration, hydrogen peroxide concentration, sulfuric acid concentration, leaching time, pulp density, and leaching temperature were optimized. The feasibility of extracting gold from the leach residue obtained from HPOL was evaluated at the defined $KI–H_2O_2–H_2SO_4$ leaching conditions under different pulp densities. The gold recovery from the pregnant leach solution via sequential precipitation using sodium hydroxide and ascorbic acid was also discussed.

2. Materials and Methods

2.1. Materials

Waste printed circuit boards (WPCBs) ash received from the company (Dowa Metals and Mining Co., Ltd., Akita, Japan) and commercially available pure gold chips (99.99%, $10 \times 10 \times 1$ mm^3 in size, AUE02CB) purchased from the chemical company (Kojunda Chemical Laboratory, Co. Ltd., Sakado, Japan) were used as starting materials in this study. The WPCBs ash was crushed and ground to pass through a 100 μm (140 mesh) sieve using a jaw crusher (Pulverisette 1, Fritsch, Fritsch GmbH & Co. KG, Idar-Oberstein, Germany) and a grinder (HERZOG, HP-M 100, Maschinenfabrik GmbH & Co. KG, Osnabrück, Germany).

The pure gold chips were first flattened and then cut into small pieces (2 × 2 × 1 mm^3) and were used in iodide leaching to optimize the process.

The chemical composition of the WPCBs ash sample was analyzed with inductively coupled plasma optical emission spectrometry (ICP-OES, SPS-5510, Seiko Instruments Inc., Tokyo, Japan) and X-ray fluorescence (XRF, ZSX Primus II, Rigaku Corporation, Tokyo, Japan). The results obtained are presented in Table 1.

Table 1. The chemical composition of the waste printed circuit boards (WPCBs) ash sample.

Main Metals and Non-Metal in WPCBs Ash, wt. %											
Au	Ag	Pd	Cu	Zn	Ni	Al	Co	Fe	Pb	Sn	Si
0.03	0.04	0.02	20.7	1.6	0.7	4.5	0.02	2.5	1.7	3.55	17.5

The particle size distribution of the sample of WPCBs ash was characterized using a particle size analyzer (Microtrac, MT3300EXII, Nikkiso Group, Osaka, Japan) (Figure 1). The main components in the sample identified using X-ray diffractometer (XRD, RINT-2200/PC, Rigaku, Tokyo, Japan) were of copper oxide (CuO), quartz (SiO_2), aluminum oxide (Al_2O_3), and tin oxide (SnO_2), as shown in Figure 2.

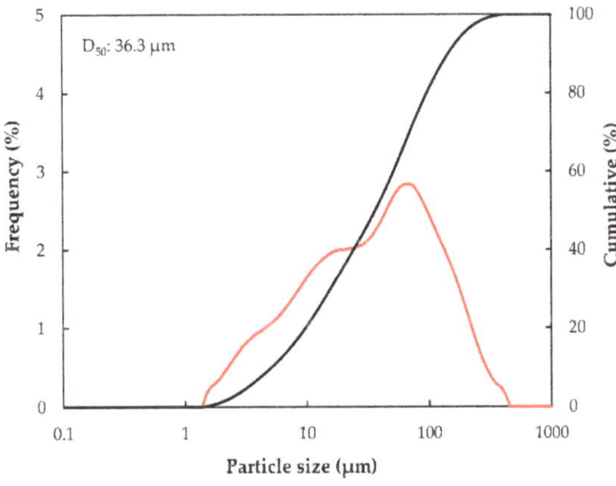

Figure 1. Particle size distribution of waste printed circuit boards (WPCBs) ash.

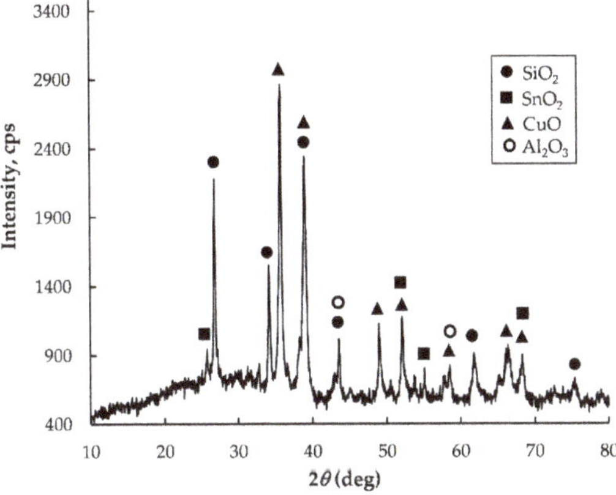

Figure 2. X-ray diffraction pattern of WPCBs ash.

Sulfuric acid (H_2SO_4), potassium iodide (KI), iodine (I), hydrogen peroxide (H_2O_2), sodium hydroxide (NaOH), and L-ascorbic acid, L-AA ($C_6H_8O_6$) were purchased from Tokyo Chemical Industry Co., Ltd. (Tokyo, Japan) and Wako Pure Chemical Industries, Ltd. (Osaka, Japan). All the chemical reagents were analytical grade and used as received. Distilled water was used to prepare all aqueous solutions: 1 M H_2SO_4, 1 M H_2O_2, 18-72 mM KI, 0.1 M NaOH, and 0.1 M L-AA. Chemical composition and constituents of the WPCB ash sample, solid residues from leaching, and precipitates were characterized using XRD, XRF, and field emission-scanning electron microscope combined with energy dispersive x-ray spectroscopy (FE/SEM-EDS), JSM-7800F (JEOL, Tokyo, Japan). Concentrations of metals in the aqueous phases from leaching and precipitation were determined using ICP-OES (Seiko Instruments Inc., Tokyo, Japan). The changes of pH and oxidation-reduction potential (ORP) of the solutions through leaching and precipitation of metals under different media were examined by pH/ORP meter (Laqua, D-74, Horiba, Ltd., Kyoto, Japan). A vacuum pump (Buchi, V-700, BUCHI Labortechnik AG, Flawil, Switzerland) and a mini centrifuge (Cat. Number C0302, Argos Technologies, Inc., Elgin, IL, USA) were used to separate solids from liquid phases after leaching and precipitation experiments.

2.2. Leaching Experiments

The process optimization, separation, and extraction of the base and precious metals from the WPCBs ash via a two-step leaching procedure involving high pressure oxidative leaching (HPOL) and iodide leaching are described in this section.

2.2.1. Optimization of Iodide Leaching

Gold leaching experiments were carried out according to the following procedure: a certain amount of small pieces ($2 \times 2 \times 1$ mm^3) of pure gold chips was inserted in a constant volume (10 mL) of an aqueous solution consisting of a mixture of iodide, hydrogen peroxide, and sulfuric acid (KI–H_2O_2–H_2SO_4) into a lidded 30 mL Teflon beaker, and placed in a glass vessel containing water onto a hot plate. In order to optimize the gold leaching in the system, various relevant variables, which are iodide concentration (18–72 mM KI), H_2SO_4 concentration (0–100 mM), H_2O_2 concentration (5–30 mM), leaching temperature (20–80 °C), and leaching time (2–12 h), were investigated under various intervals at stirring speed of 550 rpm.

After iodide leaching, undissolved pure gold species were separated from the liquor solution by filtration and dried into a drying oven at 80 °C for 24 h. The efficiency of gold extraction was calculated by weighing the amount of gold undissolved and comparing the weight loss of gold to the initial gold weight used in a leaching experiment. The liquor solution was analyzed by ICP-OES (Seiko Instruments Inc., Tokyo, Japan) for determination of gold concentration.

2.2.2. High Pressure Oxidative Leaching

Leaching experiments were carried out in an autoclave equipped with a 200 mL Teflon vessel (Nitto High Voltage Co., Ltd., Tsukuba, Japan), impeller, inlet, and outlet gas pipes, pressure gauge and temperature sensor, and electrical heating system (Figure S1). An amount of 10 g of a powder sample of WPCBs ash was firstly mixed with 100 mL, 1 M H_2SO_4 solution into a vessel, and placed the prepared slurry into the autoclave. Then, the autoclave system was heated up to the selected temperature (100–180 °C) and set to an impeller speed of 750 rpm. After that, pure oxygen gas was injected to the slurry into the vessel by adjusting the total pressure (P_{tot}) of 2 MPa using a pressure gauge and held for 30 min. The total pressure (P_{tot}) includes a partial pressure of oxygen gas (P_{ox}) and vapor pressure (P_v) into the vessel in the autoclave. After the leaching experiment, the autoclave was cooled down up to room temperature, and the slurry from the leaching was filtered using a membrane filter (Toyo Roshi Kaisha, Ltd., Tokyo, Japan) by a vacuum pump. The composition of leachate and solid residue generated from HPOL were analyzed by various techniques to estimate the extraction efficiency of metals.

2.2.3. Iodide Leaching for Extraction of Gold from WPCBs Ash

The leach residue of WPCBs ash obtained from HPOL was used as a sample to extract precious metals especially gold via iodide leaching. The leaching experiments were conducted at varying pulp densities ranging from 50 g/L to 200 g/L under the optimized conditions explained in Section 2.2.1 (72 mM KI, 20 mM H_2SO_4, 15 mM H_2O_2, stirring speed of 550 rpm, at 60 °C for 8 h). After completing the leaching period, solid and liquid phases were separated using the filtration with the membrane filter. The solid residue was analyzed by XRD and XRF, respectively. The concentrations of metals in a pregnant leach solution were determined by ICP-OES. The efficiencies of metals dissolution in the iodide solution and accompanying metals remained in the residue were calculated by mass balance.

2.3. Evaluation of Effectiveness of Iodide Leaching

The extraction of precious and base metals from the residue of WPCBs ash via iodide ($KI–H_2O_2–H_2SO_4$) leaching was evaluated with the comparison of iodine–iodide ($I–KI–H_2O$) leaching. The experiments were carried out using 20 mM H_2SO_4, 15 mM H_2O_2 in the iodide solution, and 0.08 mM I_2 in the iodine–iodide solution under the same condition of KI concentration (72 mM), pulp density of 50 g/L, temperature (60 °C), leaching time (8 h), and stirring speed (550 rpm).

2.4. Precipitation of Gold from Pregnant Leach Solution

The precipitation experiments were conducted in 20 mL of plastic tube at room temperature (25 °C) at a stirring speed of 550 rpm for 10 min. Firstly, minor amounts of metal impurities in the pregnant leach solution were selectively removed by precipitation under the alkaline conditions using a 0.1 M NaOH solution. Then, gold that remained in the resulting alkaline solution was reduced into its elemental form in acidic conditions using 0.1 M L-AA, which is known as vitamin C as a dietary supplement. After the precipitation experiment, a colloidal solution centrifuged at 9200 rpm for 5 min and the precipitate obtained was washed with 5 mL distilled water and then dried at 80 °C for 24 h. The precipitates were characterized by SEM-EDS. To determine the efficiency of metal precipitation, the supernatant solution from the centrifugation was analyzed by ICP-OES.

3. Results and Discussion

3.1. Optimization of Iodide Leaching

To properly and better optimize the iodide leaching process, pure gold chips dissolved in a mixture composed of potassium iodide, hydrogen peroxide, and sulfuric acid ($KI–H_2O_2–H_2SO_4$), preventing an interference effect of accompanying metals existing in WPCBs on gold extraction.

The generation of iodine species via oxidation of iodide with hydrogen peroxide in acidic media and the dissolution of gold in the solution are represented by the following Equations (1)–(5) [27–31,36,37]:

$$2I^- + H_2O_2 + 2H^+ \rightarrow I_2 + 2H_2O \qquad E = 1.143\ V \tag{1}$$

$$I_2 + I^- \rightarrow I_3^- \qquad K = 713 \tag{2}$$

$$Au + 2I^- \rightarrow AuI_2^- + e^- \qquad E = -0.578\ V \tag{3}$$

$$Au + 4I_3^- \rightarrow 3AuI_4^- + e^- \qquad E = -0.757\ V \tag{4}$$

$$2Au + I^- + I_3^- \rightarrow 2AuI_2^- \qquad E = -0.042\ V \tag{5}$$

Leaching parameters such as potassium iodide concentration, hydrogen peroxide concentration, sulfuric acid concentration, leaching temperature, and leaching time that affect the dissolution of gold in the $KI–H_2O_2–H_2SO_4$ system were investigated, and results obtained are summarized in the following sections.

3.1.1. Effect of Potassium Iodide Concentration and Hydrogen Peroxide Concentration

The gold extraction was examined under different concentrations of KI (18–72 mM) and H_2O_2 (5–30 mM) in dilute aqueous solutions of H_2SO_4 (40 mM). The leaching experiments were performed at 40 °C, 550 rpm for 2 h. The effects of KI and H_2O_2 concentrations on gold extraction from the gold chips were exhibited in Figure 3. Results showed that gold extraction greatly varied with variations in the concentration of KI and H_2O_2, respectively. At 18 mM KI concentration, gold extraction decreased greatly from 84.2 mg/L to 0.2 mg/L as H_2O_2 concentration increases from 5 to 30 mM. Whereas at KI concentrations of 48 mM and 72 mM, rapid increases in the gold concentration from 113.3 mg/L to 180.7 mg/L and 166.5 mg/L to 279 mg/L were observed with the addition of 5–10 mM and 5–15 mM H_2O_2, respectively. When the H_2O_2 concentration increased further up to 30 mM, the extraction of gold decreased to 34.2 mg/L and 112.8 mg/L, respectively. It appears quite clearly in Figure 3 that the maximum gold concentrations of 84.2 mg/L, 180.7 mg/L, and 279 mg/L were achieved after 2 h leaching at 40 °C in KI–H_2O_2–H_2SO_4 solutions composed of 18 mM KI, 40 mM H_2SO_4, and 5 mM H_2O_2; 48 mM KI, 40 mM H_2SO_4, and 10 mM H_2O_2; and 72 mM KI, 40 mM H_2SO_4, and 15 mM H_2O_2, respectively. It implies that the highest gold extraction achieved with KI–H_2O_2 molar ratios within the range of 4:1–5:1. As the molar ratio of KI–H_2O_2 was less than 3:1, the gold extraction was decreased. Results indicate that the appropriate molar amounts of KI and H_2O_2 in the leaching mixture is the most important to generate a suitable amount of tri-iodine species (I_3^-) that promotes the gold extraction from the gold sample. It suggests that the deficient amounts of H_2O_2 (5–10 mM) lead to produce insufficient concentrations of I_3^-; on the contrary, the excessive amounts of H_2O_2 (30 mM) result in the formation of solid iodine ($I_{2(s)}$) under oxidizing conditions as indicated by Equation (6):

$$2I^-_{(aq)} + H_2O_{2(aq)} + H^+_{(aq)} \rightarrow I_{2(s)} + 2H_2O_{(l)} \quad \text{(high oxidation condition)} \quad (6)$$

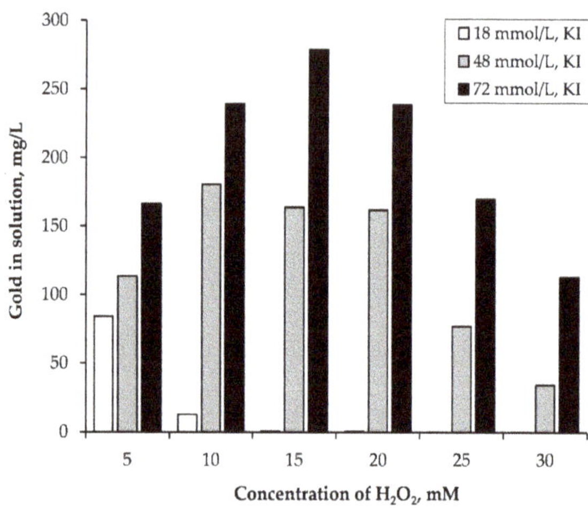

Figure 3. Gold extraction as functions of KI and H_2O_2 concentrations in the KI–H_2O_2–H_2SO_4 solution.

As a result, 72 mM KI and 15 mM H_2O_2 were selected as the suitable molar amount (5:1) between KI and H_2O_2 for the subsequent experiments.

3.1.2. Effect of Sulfuric Acid Concentration

Figure 4 shows the effect of sulfuric acid concentration on the gold extraction from pure gold chips. The concentration of H_2SO_4 in the leaching solution consisting of 72 mM KI and 15 mM H_2O_2

was varied between 0 (without H_2SO_4) and 100 mM. Other experimental parameters were fixed as described above.

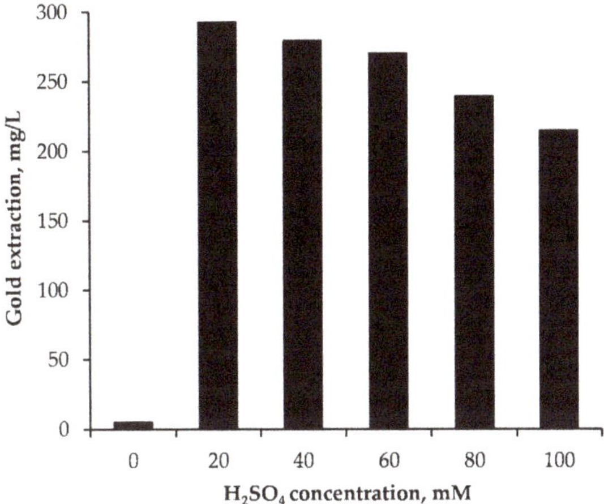

Figure 4. Gold extraction as a function of H_2SO_4 concentration in the KI–H_2O_2–H_2SO_4 solution.

Results indicate that the absence of H_2SO_4 in the leaching solution results in lower gold extraction (5 mg/L), whereas the presence of H_2SO_4 leads to a positive effect on the gold extraction. At 20 mM H_2SO_4 concentration, gold extraction increased significantly to a maximum of 292 mg/L Au. Whereas an increase in H_2SO_4 concentration further up to 100 mM resulted in a decrease in Au concentration to 214.6 mg/L. It suggests that the efficiency of the gold extraction depends on the molar ratio of the reactants in the leaching solution. It is estimated that the molar ratios of KI–H_2O_2 –H_2SO_4 in the solutions were in the range between 5:1:0 and 5:1:7. The maximum gold extraction of 292 mg/L was achieved at the molar ratio between KI–H_2O_2–H_2SO_4 of 5:1:1. The decrease in gold extraction might be explained by the formation of a precipitate of iodine ($I_{2(s)}$) due to the presence of 40–100 mM H_2SO_4, at which the values of pH and Eh of solutions did not change obviously, whereas these values changed drastically with 20 mM H_2SO_4 (Figure 5).

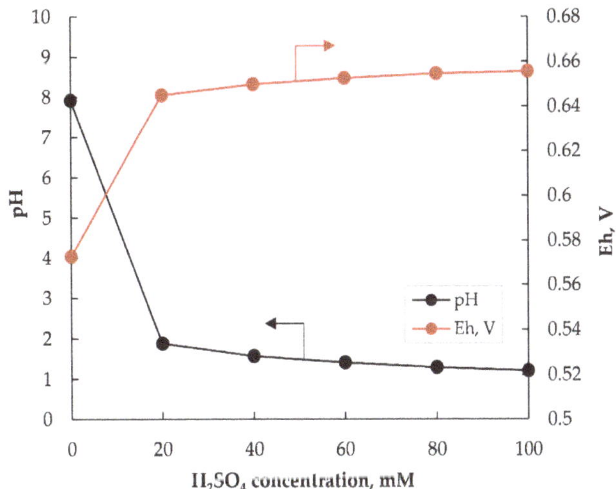

Figure 5. Effect of H_2SO_4 concentration on pH and Eh in the leaching solution.

As a result, the most appropriate sulfuric acid concentration for the extraction of gold was selected as 20 mM for the next experiments.

3.1.3. Effect of Leaching Temperature and Time

The effect of leaching temperature and leaching time on gold extraction from the gold chips were investigated under different temperature ranges from 20 °C to 80 °C and the various times from 0.25 h to 12 h. Other experimental variables such as KI concentration, H_2O_2 concentration, H_2SO_4 concentration, and stirring speed were kept constant explained in the above experiments. The results are summarized in Figure 6. It showed that the extraction of gold rises dramatically with an increase in leaching temperature and time, respectively. At the temperatures of 20 °C and 40 °C, gold extraction grew continuously for a period of leaching time (0.2–12 h), and reached its highest levels of 670 mg/L and 1127 mg/L from 18 mg/L and 30 mg/L, respectively. The maximum extraction of gold was 1137 mg/L at 60 °C after 8 h leaching that was equal to 99% extraction efficiency of gold. Whereas at 80 °C, the most effective extraction of gold had achieved 1064 mg/L for 6 h leaching that was consistent with 92.6% extraction efficiency of gold. Further increase in the temperature and time results in the decrease in the extraction of gold (Figure 6). The results revealed that the gold extraction with the KI–H_2O_2–H_2SO_4 solution is dependent on various leaching parameters such as iodide concentration, hydrogen peroxide concentration, sulfuric acid concentration, leaching temperature, and leaching time, respectively. Consequently, the best conditions for gold leaching in the KI–H_2O_2–H_2SO_4 solution determined were: KI concentration of 72 mM, H_2O_2 concentration of 15 mM, H_2SO_4 concentration of 20 mM, stirring speed of 550 rpm, leaching temperature of 60 °C, and leaching time of 8 h.

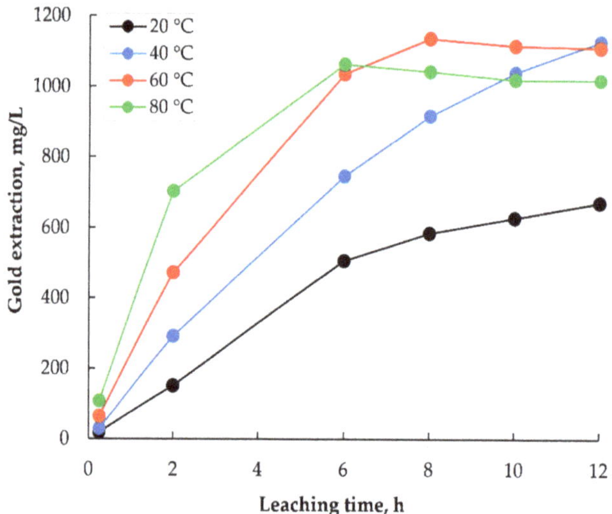

Figure 6. Gold extraction as a function of leaching time under different temperatures.

3.2. *Extraction of Base Metals from WPCBs Ash by HPOL*

The presence of higher amounts of metal impurities in WPCBs causes adverse effects on precious metals extraction and reagent consumption [30]. To reduce the side effects of metal impurities on gold extraction and iodide consumption, HPOL experiments were carried out at different temperatures.

Effect of Temperature

High pressure oxidative leaching as a pretreatment for recycling of base metals from the WPCBs ash was conducted at different temperatures ranging from 100 to 180 °C using a laboratory-scale autoclave under conditions of fixed sulfuric acid concentration (1 M), pulp density (10 g/t), agitation speed (750 rpm), total pressure (2 MPa), and leaching time (30 min). Figure 7 shows the extraction percentage of base metals from the WPCBs ash under the conditions. Results showed that the maximum extraction of Cu (99%), Zn (95.7%), Ni (91%), Al (87.3%), Co (82%), and Fe (70%), which are main base metals in the pulverized WPCBs ash, were achieved in 1 M H_2SO_4 at 160 °C, while only 1.3% of Pb was dissolved. The reaction between metal oxides and a dilute acid is often quite slow because the oxidant (H^+, SO_4^{2-}) is not enough to oxidize the metal. However, it is possible to take place the progress of the reaction in the presence of oxygen that acts as an oxidizing agent. The dissolution of base metals with dilute sulfuric acid in the presence of oxygen is, hence, represented as follows [38,39]:

$$CuO + H_2SO_4 \rightarrow CuSO_4 + H_2O \tag{7}$$

$$ZnO + H_2SO_4 \rightarrow ZnSO_4 + H_2O \tag{8}$$

$$CoO + H_2SO_4 \rightarrow CoSO_4 + H_2O \tag{9}$$

$$NiO + H_2SO_4 \rightarrow NiSO_4 + H_2O \tag{10}$$

$$Al_2O_3 + 3H_2SO_4 \rightarrow Al_2(SO_4)_3 + 3H_2O \tag{11}$$

$$Fe_2O_3 + 3H_2SO_4 \rightarrow Fe_2(SO_4)_3 + 3H_2O \tag{12}$$

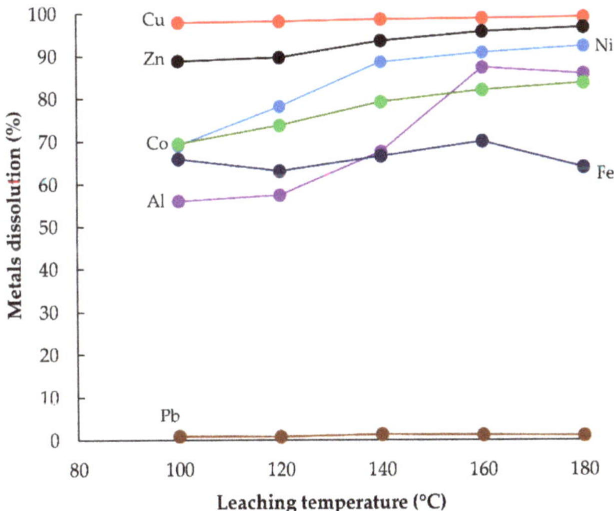

Figure 7. Extraction percentage of base metals as a function of temperature. (Conditions: 1 M H_2SO_4, 100–180 °C, 10 g/t pulp density, 750 rpm stirring speed, 2 MPa pressure for 30 min leaching).

It can be seen that precious metals (Au, Ag, and Pd) and tin (Sn) in the WPCBs ash were not dissolved in 1 M H_2SO_4 under the high pressure oxidation conditions due to their extremely slow leaching kinetics. It may require strong H_2SO_4 and longer leaching time to extract these metals.

The chemical composition of the leachate and the solid residue resulted from HPOL with 1 M H_2SO_4 are presented in Table 2.

Table 2. The chemical composition of the leachate and the solid residue of the WPCBs ash after high-pressure oxidative leaching (HPOL).

Sample	Au	Ag	Pd	Cu	Zn	Ni	Al	Co	Fe	Pb	Sn	Si
Leachate, (g/L)	0	0	0	22.0	1.6	0.06	3.4	0.02	1.7	0.01	0	0
Leach residue, (wt. %)	0.06	0.21	0.02	0.23	0.07	0.04	0.48	0.004	0.7	1.67	7.7	24.4

It shows that the concentration of precious metals in the leach residue was higher than that in the WPCBs ash. It is worth noting that the residue is a potential source of precious metals. On the other hand, the leachate contained the proper amount of base metals for recovery. It suggests that HPOL could be used to concentrate base metals in the leachate and precious metals in the leach residue, separately. Therefore, the leach residue obtained by HPOL was used to recover precious metals by iodide leaching in a further study.

The intensities of diffraction peaks corresponding to the copper phases in the residue became rather low compared to those from the initial WPCBs ash sample (Figure S2).

3.3. *Extraction of Precious Metals from Leach Residue via Iodide Leaching*

Effect of Pulp Density

The leach residue of WPCBs ash obtained from HPOL was dissolved in a mixture of $KI–H_2O_2–H_2SO_4$ under the selected conditions described in the previous section (Section 3.1). Figure 8 shows the percentage of metals extracted from the leach residue of WPCBs under the different pulp densities of 50 g/t, 100 g/t, and 200 g/t at the optimum conditions. The results show that an increase in the pulp density leads to a decrease in the percentage extraction of metals from the leach residue. It means that the extraction of metals had similar tendencies under the conditions.

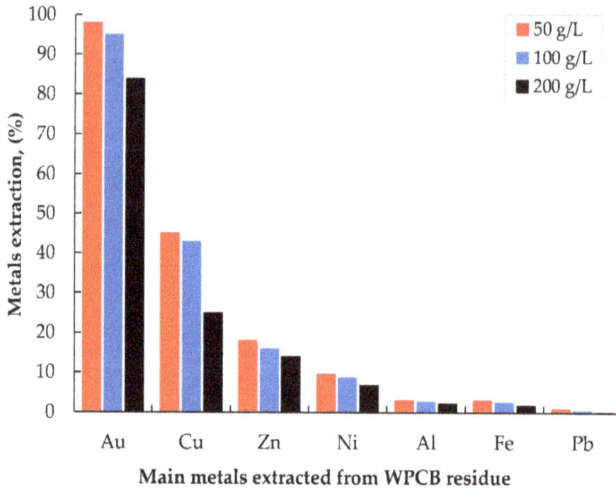

Figure 8. The percentage extraction of metals as a function of pulp density.

It is observed that gold and copper are preferentially extracted from the residue via $KI–H_2O_2–H_2SO_4$ leaching. The percentage extraction of gold and copper decreased from 98% to 84% and from 45% to 25%, respectively, with an increase in the pulp density, whereas the percentages of both silver and palladium extraction were lower than 1%. However, the percentage extraction of metals at high pulp density reduced obviously than that in lower pulp densities, and the concentrations of metals in pregnant leach solutions increase with an increase in pulp density, as shown in Table 3.

Table 3. Concentration of metals in the pregnant leach solutions after leaching, mg/L.

Pulp Density, g/L	Au	Cu	Zn	Ni	Al	Fe	Pb
50	32.3	17.2	11.5	8.2	29.2	19.0	14.0
100	58.2	34.2	24.7	18.0	64.4	41.2	13.2
200	108	40.7	43.4	26.6	102.8	55.7	9.2

3.4. Evaluation of Effectiveness of Iodide Leaching

Figure 9 presents the results of a quantitative comparison of the effectiveness of the iodide ($KI–H_2O_2–H_2SO_4$) and iodine–iodide ($I_2–KI–H_2O$) leaching for precious and base metals extraction from the leach residue of WPCBs ash under the same leaching conditions.

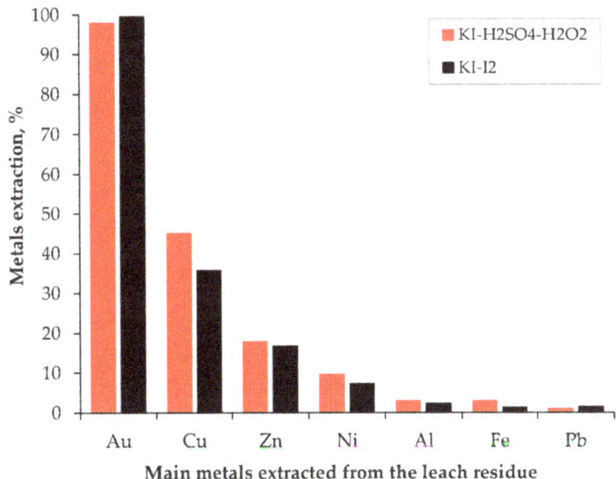

Figure 9. Comparison of the efficiency of metals extraction via iodide leaching and iodine–iodide leaching at pulp density of 50 g/L.

The results show that the extraction of gold and lead by leaching with $KI–I_2–H_2O$ are somewhat higher than those achieved with $KI–H_2O_2–H_2SO_4$. On the contrary, the efficiencies of other metals (Cu, Zn, Ni, Al, and Fe) extraction with $KI–H_2O_2–H_2SO_4$ were slightly higher than that in $KI–I_2–H_2O$ due to the existence of SO_4^- ions in $KI–H_2O_2–H_2SO_4$ leaching system. It was observed that no significant differences resulted in the gold extraction from the sample by both iodide and iodine–iodide leaching. It suggests that iodide leaching is as much an effective method as iodine–iodide leaching for the extraction of gold from samples containing gold and can potentially reduce reagent cost, particularly in iodine consumption.

3.5. Recovery of Gold from the Pregnant Leach Solution

The pregnant leach solution (PLS) containing 32.3 mg/L Au and considerable amounts of base metal impurities such as Cu, Zn, Ni, Al, Fe, and Pb was used in the subsequent study (Table 3). The PLS has a pH of 1.9 and Eh of 0.64 V. The recovery of gold from the PLS consists of two steps: (1) removal of metal impurities via precipitation using NaOH; (2) gold recovery from the resulting solution by reductive precipitation using L-AA as a reducing agent. The results of the sequential precipitation are presented in the following sections.

3.5.1. Removal of Metal Impurities from the PLS

Based on the preliminary experimental results [25,40], the pH of the PLS was adjusted up to 9.0 from 1.9 using 0.1 M NaOH at ambient temperature and stirring speed of 500 rpm. The efficiency of metals precipitation is shown in Figure 10. The results show that the vast majority of Cu (99.6%), Zn (100%), Ni (90.4%), Al (100%), Fe (100%), and Pb (97%) as metal impurities were precipitated from the PLS, whereas gold did not precipitate under the condition. It indicates that most gold (32 mg/L) and trace amounts of Cu (618 µg/L), Ni (665.6 µg/L), and Pb (237.3 µg/L) have remained into the resulting solution which had a pH of 9.0 and Eh of 0.21 V. It implies that this process is efficient in removal of metal impurities from PLS.

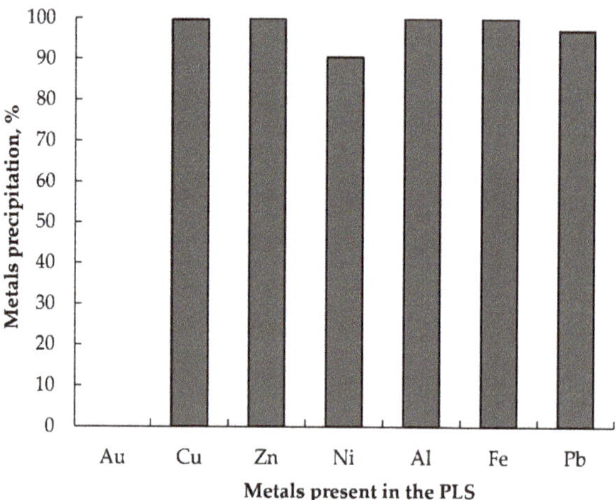

Figure 10. Removal of metal impurities from the pregnant leach solution (PLS) by precipitation at pH 9.

3.5.2. Recovery of Gold from the Solution by Reductive Precipitation Using L-AA

Metals remained in the alkaline solution that resulted from the precipitation of base metals with NaOH were recovered by reductive precipitation using a 0.1 M, L-AA solution under conditions that were fixed as the molar ratio between Au and L-AA of 1:1, temperature of 25 °C, stirring speed of 500 rpm and precipitation time for 10 min [25,40]. Figure 11 shows the recovery of metals like Au, Cu, Ni, and Pb from the solution via precipitation with L-AA.

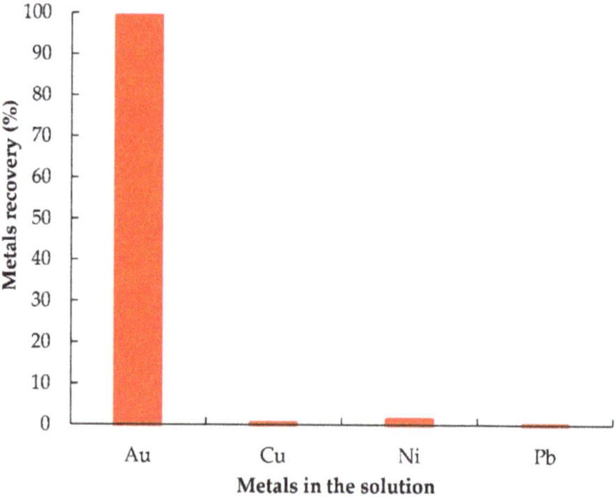

Figure 11. Recovery of metals from the solution by reductive precipitation with L-AA.

The result showed that gold recovery of 99.2% was achieved from the solution via the precipitation with L-AA under the conditions, while yields of Cu (0.4%), Ni (1.4%), and Pb (0.1%) as metal impurities in the precipitate were negligible. It was observed that the addition of L-AA results in a reduction of pH of the solution from 9 to 1.8 and an increase of Eh from 0.21 V to 0.64 V. The Eh-pH diagram for gold iodide and ascorbic acid species in the solution was constructed at 25 °C within the pH range from 0 to 14 using STABCAL software (N NBS, Helgeson thermodynamic data source, W32-Stabcal Version 1.0, Montana Technological University (Montana Tech), Butte, MT, United States) as shown in Figure 12.

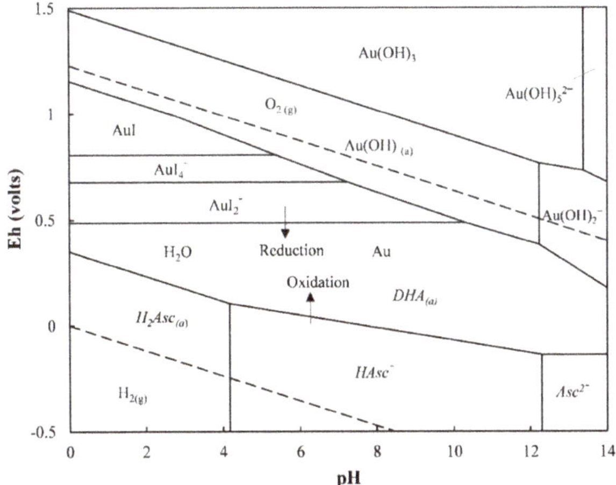

Figure 12. The Eh–pH diagram of gold–iodide and ascorbic acid species in the solution. (Condition: [Au] = 32.3 mg/L, [I] = 0.072 M (72 mM), Eh = −0.5–1.5 volts, *P* = 1 atm at pH 0–14, 25 °C, (NBS, STABCAL software)).

The thermodynamic data for gold iodide and L-AA related species were imported from the references [25,41–43]. Table S1 summarizes the thermodynamic data for individual species in the solution used to construct the Eh–pH diagram. As shown in Figure 12, L-AA/H_2Asc is converted into its oxidized forms such as $HAsc^-$, Asc^{2-}, and DHA at the different pH conditions as indicated by Equation (13) [43–45].

$$H_2Asc \rightleftharpoons HAsc^- \rightleftharpoons Asc^{2-} \rightleftharpoons DHA \quad (13)$$

It implies that DHA-dehydroascorbic acid ($C_6H_6O_6$) is the most stable oxidized product of L-AA in the pH range from 0 to 14.

These findings suggest that colloidal gold can be obtained from gold–iodide solutions by reductive precipitation with L-AA. The mechanism of the reduction of gold iodide complexes via L-AA is generally explained by the Equations (14)–(16) [25,46]:

$$C_6H_8O_6 = C_6H_6O_6 + 2H^+ + 2\bar{e} \qquad E = 0.35\ V \quad (14)$$

$$AuI_4^- + 2\bar{e} = AuI_2^- + 2I^- \qquad E = 0.551\ V \quad (15)$$

$$AuI_2^- + \bar{e} = Au^0 + 2I^- \qquad E = 0.578\ V \quad (16)$$

The colloidal gold as a precipitate formed from the reductive precipitation was analyzed by using FE/SEM-EDS. The SEM result shown in Figure S3 indicates that produced particles do not have a definite shape and size due to the agglomeration of particles formed. The EDS analysis in the area identified a sharp peak corresponding to gold (Figure S4).

4. Conclusions

This study aimed to recover gold from WPCBs ash by hydrometallurgical processes involving HPOL and potassium iodide leaching (KI–H_2O_2–H_2SO_4) followed by sequential precipitation using NaOH and L-AA. The following conclusions can be drawn based on the results obtained from this study:

- The maximum gold extraction of 1137 mg/L (99% extraction efficiency) was achieved from pure gold chips in KI–H_2O_2–H_2SO_4 media under the conditions determined through this study, while the molar ratio of KI–H_2O_2–H_2SO_4 reactants was 5:1:1.
- The most effective extraction of Cu (99%), Zn (95.7%), Ni (91%), Al (87.3%), Co (82%), and Fe (70%) from the WPCBs ash were achieved in 1 M H_2SO_4 solution by HPOL pre-treatment at the defined conditions.
- Over 95% of gold could be extracted from the leach residue from HPOL via iodide leaching under the conditions determined from the leaching of gold chips in the KI–H_2O_2–H_2SO_4 solution.
- Results suggested that iodide leaching is an effective method for the extraction of gold from WPCBs.
- Efficient gold recovery (>99.2%) could be achieved from the PLS by reductive precipitation using L-AA after selectively removed metal impurities via precipitation with NaOH at a pH of 9.

As a result, a multi-stage hydrometallurgical procedure with greater than 93% gold recovery from WPCBs is proposed.

Supplementary Materials: The following are available online at http://www.mdpi.com/2075-4701/9/3/363/s1, Figure S1: A schematic diagram of an autoclave used in this study. Figure S2: A comparison of the XRD pattern of the leach residue of HPOL with WPCBs ash sample. Figure S3: FE/SEM image of gold particles precipitated from the solution by reductive precipitation with L-AA. Figure S4. EDS spectra of gold particles precipitated from the solution by reductive precipitation with L-AA. Table S1: The thermodynamic data for individual species for the construction of an Eh-pH diagram.

Author Contributions: Conceptualization, A.B. and A.S.; Methodology, A.B.; Software, A.B. and H.-H.H.; Validation, A.B., K.H., H.-H.H., and A.S.; Formal analysis, A.B. and K.H.; Investigation, A.B. and K.H.; Resources, A.S.; Data curation, A.B. and K.H.; Writing—original draft preparation, A.B.; Writing—review and editing, A.B., K.H. and A.S.; Visualization, A.B.; Supervision, A.B., K.H., and A.S.; Project administration, A.S.; Funding acquisition, A.S.

Funding: This research was funded by JSPS KAKENHI, Grant Number 16H04182, Japan and The APC was funded by 16H04182.

Acknowledgments: This work was mainly supported by the JSPS KAKENHI Grant Number 16H04182 and partially supported by the Program for Leading Graduate Schools, New Frontier Leader Program for Rare-Metals and Resources, JSPS. The authors gratefully acknowledge their financial support.

Conflicts of Interest: The authors declare no conflict of interest.

References

1. Jha, M.K.; Kumari, A.; Panda, R.; Kumar, J.R.; Yoo, K.; Lee, J.Y. Review on hydrometallurgical recovery of rare earth metals. *Hydrometallurgy* **2016**, *165*, 2–26. [CrossRef]
2. Li, H.; Eksteen, J.; Oraby, E. Hydrometallurgical recovery of metals from waste printed circuit boards (WPCBs): Current status and perspectives—A review. *Resour. Conserv. Recycl.* **2018**, *139*, 122–139. [CrossRef]
3. Masloboev, V.A.; Seleznev, S.G.; Svetlov, A.V.; Makarov, D.V. Hydrometallurgical Processing of Low-Grade Sulfide Ore and Mine Waste in the Arctic Regions: Perspectives and Challenges. *Minerals* **2018**, *8*, 436. [CrossRef]
4. Choubey, P.K.; Lee, J.; Kim, M.; Kim, H. Conversion of chalcopyrite to copper oxide in hypochlorite solution for selective leaching of copper in dilute sulfuric acid solution. *Hydrometallurgy* **2018**, *178*, 224–230. [CrossRef]
5. Battsengel, A.; Batnasan, A.; Narankhuu, A.; Haga, K.; Watanabe, Y.; Shibayama, A. Recovery of light and heavy rare earth elements from apatite ore using sulphuric acid leaching, solvent extraction and precipitation. *Hydrometallurgy* **2018**, *179*, 100–109. [CrossRef]

6. Priya, A.; Hait, S. Comprehensive characterization of printed circuit boards of various end-of-life electrical and electronic equipment for beneficiation investigation. *Waste Manag.* **2018**, *75*, 103–123. [CrossRef]
7. Cucchiella, F.; D'Adamo, I.; Lenny Koh, S.C.; Rosa, P. Recycling of WEEEs: An economic assessment of present and future e-waste streams. *Renew. Sust. Energ. Rev.* **2015**, *51*, 263–272. [CrossRef]
8. Holgersson, S.; Steenari, B.-M.; Björkman, M.; Cullbrand, K. Analysis of the metal content of small-size Waste Electric and Electronic Equipment (WEEE) printed circuit boards—Part 1: Internet routers, mobile phones and smartphones. *Resour. Conserv. Recycl.* **2018**, *133*, 300–308. [CrossRef]
9. Yamane, L.H.; Moraes, V.T.; Espinosa, D.C.R.; Tenório, J.A.S. Recycling of WEEE: Characterization of spent printed circuit boards from mobile phones and computers. *Waste Manag.* **2011**, *31*, 2553–2558. [CrossRef] [PubMed]
10. Işıldar, A.; Rene, E.R.; Hullebusch, E.D.; Lens, P.N.L. Electronic waste as a secondary source of critical metals: Management and recovery technologies. *Resour. Conserv. Recycl.* **2018**, *135*, 296–312. [CrossRef]
11. Hadi, P.; Xu, M.; Lin, C.S.K.; Hui, C.W.; McKay, G. Waste printed circuit board recycling techniques and product utilization. *J. Hazard. Mater.* **2015**, *283*, 234–243. [CrossRef] [PubMed]
12. Nekouei, R.K.; Pahlevani, F.; Rajarao, R.; Golmohammadzadeh, R.; Sahajwalla, V. Two-step pre-processing enrichment of waste printed circuit boards: Mechanical milling and physical separation. *J. Clean. Prod.* **2018**, *184*, 1113–1124. [CrossRef]
13. Işıldar, A.; Hullebusch, E.D.; Lenz, M.; Laing, G.D.; Marra, A.; Cesaro, A.; Panda, S.; Akcil, A.; Kucuker, M.A.; Kuchta, K. Biotechnological strategies for the recovery of valuable and critical raw materials from waste electrical and electronic equipment (WEEE)—A review. *J. Hazard. Mater.* **2019**, *362*, 467–481. [CrossRef]
14. Wang, H.; Zhang, S.; Li, B.; Pan, D.; Wu, Y.; Zuo, T. Recovery of waste printed circuit boards through pyrometallurgical processing: A review. *Resour. Conserv. Recycl.* **2017**, *126*, 209–218. [CrossRef]
15. Tuncuk, A.; Stazi, V.; Akcil, A.; Yazici, E.Y.; Deveci, H. Aqueous metal recovery techniques from e-scrap: Hydrometallurgy in recycling. *Miner. Eng.* **2012**, *25*, 28–37. [CrossRef]
16. Jung, M.; Yoo, K.; Alorro, R.D. Dismantling of Electric and Electronic Components from Waste Printed Circuit Boards by Hydrochloric Acid Leaching with Stannic Ions. *Mater. Trans.* **2017**, *58*, 1076–1080. [CrossRef]
17. Fleming, C.A. Hydrometallurgy of precious metals recovery. *Hydrometallurgy* **1992**, *30*, 127–162. [CrossRef]
18. Anderson, C.G. Alkaline sulfide gold leaching kinetics. *Miner. Eng.* **2016**, *92*, 248–256. [CrossRef]
19. Aylmore, M.G. Alternative lixiviants to cyanide for leaching gold ores. In *Advances in Gold Ore Processing*, 1st ed.; Adams, M.D., Wills, B.A., Eds.; Elsevier: Amsterdam, The Netherlands, 2005; Volume 15, pp. 501–539.
20. Xu, B.; Kong, W.; Li, Q.; Yang, Y.; Jiang, T.; Liu, X. A review of thiosulfate leaching of gold: Focus on thiosulfate consumption and gold recovery from pregnant solution. *Metals* **2017**, *7*, 222. [CrossRef]
21. Baghalha, M. The leaching kinetics of an oxide gold ore with iodide/iodine solutions. *Hydrometallurgy* **2012**, *113–114*, 42–50. [CrossRef]
22. Sousa, R.; Futuro, A.; Fiúza, A.; Vila, M.C.; Dinis, M.L. Bromine leaching as an alternative method for gold dissolution. *Miner. Eng.* **2018**, *118*, 16–23. [CrossRef]
23. Wang, H.; Sun, C.; Li, S.; Fu, P.; Song, Y.; Li, L.; Xie, W. Study on gold concentrate leaching by iodine-iodide. *Int. J. Miner. Metall. Mater.* **2013**, *20*, 323–328. [CrossRef]
24. Konyratbekova, S.S.; Baikonurova, A.; Akcil, A. Non-cyanide leaching processes in gold hydrometallurgy and iodine-iodide applications: A review. *Miner. Process. Extr. Metall. Rev.* **2015**, *36*, 198–212. [CrossRef]
25. Altansukh, B.; Haga, K.; Huang, H.H.; Shibayama, A. Gold Recovery from Waste Printed Circuit Boards by Advanced Hydrometallurgical Processing. *Mater. Trans.* **2019**, *60*, 287–296. [CrossRef]
26. Homick, R.P.; Slaon, H. Gold Reclamation Process. U.S. Patent US3,957,505, 18 May 1976.
27. Qi, P.H.; Hiskey, J.B. Dissolution kinetics of gold in iodide solutions. *Hydrometallurgy* **1991**, *27*, 47–62. [CrossRef]
28. Angelidis, T.N.; Kydros, K.A.; Matis, K.A. A fundamental rotating disk study of gold dissolution in iodine-iodide solutions. *Hydrometallurgy* **1993**, *34*, 49–64. [CrossRef]
29. Altansukh, B.; Haga, K.; Ariunbolor, N.; Kawamura, S.; Shibayama, A. Leaching and Adsorption of Gold from Waste Printed Circuit Boards using Iodine-Iodide Solution and Activated Carbon. *Eng. J.* **2016**, *20*, 29–40. [CrossRef]
30. Davis, A.; Tran, T.; Young, D.R. Solution chemistry of iodide leaching of gold. *Hydrometallurgy* **1993**, *32*, 143–159. [CrossRef]

31. Meng, X.; Han, K.N. Dissolution kinetics of gold in moderate aqueous potassium iodide solutions with oxygen under pressure. *Min. Metall. Explor.* **1997**, *14*, 1–9. [CrossRef]
32. Karrech, A.; Attar, M.; Oraby, E.A.; Eksteen, J.J.; Elchalakani, M.; Seibi, A.C. Modelling of multicomponent reactive transport in finite columns—Application to gold recovery using iodide ligands. *Hydrometallurgy* **2018**, *178*, 43–53. [CrossRef]
33. Angelidis, T.N.; Kydros, K.A. Separation of Gold from Iodine-Iodide Solutions by Cementation on Zinc Particles. *Sep. Sci. Technol.* **1996**, *31*, 1105–1121. [CrossRef]
34. Teirlinck, P.A.M.; Petersen, F.W. The nature of gold-iodide adsorption onto coconut-shell carbon. *Miner. Eng.* **1996**, *9*, 923–930. [CrossRef]
35. Zhang, H.; Jeffery, C.A.; Jeffrey, M.I. Ion exchange recovery of gold from iodine-iodide solutions. *Hydrometallurgy* **2012**, *125–126*, 69–75. [CrossRef]
36. Wong, G.T.F.; Zhang, L. The kinetics of the reactions between iodide and hydrogen peroxide in seawater. *Mar. Chem.* **2008**, *111*, 22–29. [CrossRef]
37. Milenković, M.C.; Stanisavljev, D.R. Role of Free Radicals in Modeling the Iodide–Peroxide Reaction Mechanism. *J. Phys. Chem. A* **2012**, *116*, 5541–5548. [CrossRef]
38. Habbache, N.; Alane, N.; Djerad, S.; Tifouti, L. Leaching of copper oxide with different acid solutions. *Chem. Eng. J.* **2009**, *152*, 503–508. [CrossRef]
39. Yoshida, T. Leaching of Zinc Oxide in Acidic Solution. *Mater. Trans.* **2003**, *44*, 2489–2493. [CrossRef]
40. Altansukh, B.; Haga, K.; Shibayama, A. Recovery of Valuable Metals from Waste Printed Circuit Boards by Using Iodine-Iodide Leaching and Precipitation. In *Rare Metal Technology 2018. TMS 2018. The Minerals, Metals & Materials Series, Proceedings of the TMS 2018, Phoenix, AZ, USA, 11–15 March 2018*; Springer: Cham, Switzerland; pp. 131–142. [CrossRef]
41. Schmid, G.M. Chapter 11: Copper, Silver, and Gold. In *Standard Potentials in Aqueous Solution*, 1st ed.; Bard, A.J., Parsons, R., Jordan, J., Eds.; Marcel Dekker: New York, NY, USA, 1985; pp. 313–320.
42. Mironov, I.V.; Belevantsev, V.I. Hydroxogold(I) complexes in aqueous Solution. *Russ. J. Inorg. Chem.* **2005**, *50*, 1210–1216.
43. Matsui, T.; Kitagawa, Y.; Okumura, M.; Shigeta, Y. Accurate standard hydrogen electrode potential and applications to the redox potentials of vitamin C and NAD/NADH. *J. Phys. Chem. A* **2015**, *119*, 369–376. [CrossRef]
44. Du, J.; Cullen, J.J.; Buettner, G.R. Ascorbic acid: Chemistry, biology and the treatment of cancer. *Biochim. Biophys. Acta* **2012**, *1826*, 443–457. [CrossRef]
45. Tu, Y.J.; Njus, D.; Schlegel, H.B. A theoretical study of ascorbic acid oxidation and $HOO^{\cdot}/O_2^{\cdot-}$ radical scavenging. *Org. Biomol. Chem.* **2017**, *15*, 4417–4431. [CrossRef] [PubMed]
46. Serpe, A.; Rigoldi, A.; Marras, C.; Artizzu, F.; Mercuri, M.L.; Deplano, P. Chameleon behaviour of iodine in recovering noble-metals from WEEE: Towards sustainability and "zero" waste. *Green Chem.* **2015**, *17*, 2208–2216. [CrossRef]

Article

Valorization of Mining Waste by Application of Innovative Thiosulphate Leaching for Gold Recovery

Stefano Ubaldini [1,*], Daniela Guglietta [1], Francesco Vegliò [2] and Veronica Giuliano [3]

[1] Institute of Environmental Geology and Geoengineering, Italian National Research Council, Area della Ricerca di Roma RM 1—Montelibretti, Via Salaria Km 29,300—Monterotondo Stazione, 00015 Roma, Italy; daniela.guglietta@igag.cnr.it

[2] Department of Industrial and Information Engineering and Economics, University of L'Aquila—via G. Gronchi 18, Zona Ind. le di Pile, 67100 L'Aquila, Italy; francesco.veglio@ing.univaq.it

[3] Department of Earth Sciences and Environmental Technologies, Italian National Research Council, 00015 Roma, Italy; veronica.giuliano@cnr.it

* Correspondence: stefano.ubaldini@igag.cnr.it; Tel.: +39-06-90672748

Received: 18 January 2019; Accepted: 23 February 2019; Published: 28 February 2019

Abstract: The metals and industrial minerals contained in the tailings of mining and quarrying activities, can degrade natural environments as well as human health. The objective of this experimental work is the application of innovative and sustainable technologies for the treatment and exploitation of mining tailings from Romania. Within this approach, the recovery of high grade raw materials to be placed on the market is achieved and reduction of these wastes volume are achieved. The current study is focused on hydrometallurgical process for the recovery of gold. The innovative treatment chosen is the thiosulphate process that, compared with the conventional cyanide, has several advantages (e.g., it is more ecologically friendly and is not toxic to humans). The conventional cyanidation process shows operating limits in the case of auriferous refractory minerals, such as Romanian wastes, the object of the study. An important characteristic of thiosulphate leaching process it has the best selectivity towards gold; it does not attack the majority of the gangue mineral constituents. Gold extraction of 75% was obtained under ambient conditions of temperature. Moreover, the overall process achieved about 65–67% Au recovery, this being in line with the conventional cyanidation process. As these results are obtained by application of the thiosulfate process on a low gold content ore, they may be considered encouraging. The optimization of process parameters and operating conditions, should permit the best results in terms of process yields to be achieved.

Keywords: mining waste; hydrometallurgical processes; leaching kinetic; thiosulphate leaching; electrowinning; gold

---

1. Introduction

Several studies have shown that mining activities can significantly damage, pollute and alter the environment (e.g., soil, sediments, water and air quality) as well as human health [1–4]. Furthermore, mining involves the creation of surface features that are unstable, prone to landslides and collapses, but, especially, expose the environment to exogenous vast areas of mineralized rocks and byproducts of mineral treatment. These outcomes produce significant changes in the chemical environment [5,6]. The large volumes of wastes produced by mining occupy huge areas; these accumulations can substantially change the original landscape.

Alongside mining, mining wastes, if managed properly, can be important to the economic growth of countries. The application of a proper treatment ensures environmental sustainability and minimize the risks to human health. In fact, the deposits of mining tailings should not be treated as inert wastes, but as a neo-mine: only a short-sighted vision would be limited solely to neutralize the harmful and toxic waste stored and to put in place procedures and interventions that allow the restoration of the original environmental conditions.

Instead, the enhancement of the tailings, in addition to the sale of raw materials grade high (precious metals such as gold), allows a complete and effective recovery of environmental conditions sustainable because the recycling activity also determines an economic convenience for the treatment.

Romania, like many other nations affected by a long history of mining, is now grappling with environmental and social issues related to tailings produced by mining activities. On the Romanian territory a total of 300 tailings deposits, produced by exploitation of different minerals, have been inventoried. All are of significant proportion and should be submitted to amelioration procedures, such as the neutralization of treatment residues [7,8]. These waste materials are a result of mining activities that have reached the end of the production cycle due to depletion of the reservoir or, after stopping of the process for environmental reasons (i.e., regulations put in place at the time of the Convention for Romania's entry into the EU). In Romania, the administration of these landfills is carried out by the Romanian National Agency for Mineral Resources (ANMR). The mining sites considered within this study are Bălan, Deva deposit 1, Deva deposit 2, Brad Ribita and Brad Criscior. Mine Bălan, for example, which ceased its operations in 2006, under the Convention for the entry of Romania into the EU, is a tabular deposit tectonized in metamorphic rocks. The main mineral is chalcopyrite. The remaining reserves amounted to 500,000 tonnes at 0.8% of copper content. These reserves are substantially unavailable. The costs of re-mining and carrying out of modifications to enrich the mineral are too high compared to the value of resources and costs of extraction [7–10].

The deposits of tailings, left on the land during the mining of copper ore, contain 24.5 million tonnes of slag and more than 10 million cubic meters of materials that constitute a toxic and harmful waste. These deposits are located at a distance of few kilometers from the mining center (at the neighborhood of the Bălan town) [7,8].

The removable high-grade raw materials, from the tailings of Bălan, like those of the other mentioned sites, are:

- industrial minerals, such as quartz, pyrite and chlorite, recoverable through metallurgical processes;
- base metals, including copper and zinc, and precious metals such as gold and silver, extracted with hydrometallurgical processes.

The aim of this work is to apply innovative technologies for the treatment and exploitation of Romanian mining tailings. In particular, this study is focused on the development of an hydrometallurgical process for the recovery of gold from solid wastes of mining industry. The innovative treatment chosen was the thiosulphate process, that has advantages over the conventional cyanide and is non-toxic to humans; in fact, it is environmental impact is lower than for cyanidation [11–18].

The kinetics of the thiosulphate process has been studied with the main aim to improve the characteristics that distinguishes from the conventional cyanide process for gold recovery. The main potential advantages of this innovative treatment for gold recovery, can be summarized in the following points:

1. solubilization of gold with an appreciable rate of dissolution;
2. ammonia-containing solutions do not attack most of the gangue mineral constituents;
3. production of leach solutions, practically free of metallic elements that hinder the subsequent recovery operations;
4. ability to recirculate the ammonia leaching solutions;
5. non-toxicity of the reagents used in the leaching process;

6. possibility for recovery of the dissolved gold by known techniques, such as carbon adsorption and electrodeposition.

The greatest criticality of the process is constituted by the chemistry of the ammonia-thiosulphate system, that is very complicated due to the simultaneous presence of complexing ligands such as ammonia and thiosulphate, the Cu(II)–Cu(I) redox couple and the possibility of oxidative decomposition reactions of thiosulphate involving the formation of additional sulphur compounds such as tetrathionate [11].

The ultimate purpose of this work is to develop a process scheme on the basis of results obtained at laboratory scale. This will be used to perform a preliminary study of the technical feasibility of process. Figure 1 shows an example of a mining site with tailings from Romania.

Figure 1. Tailings of a mining site in Romania.

2. Materials and Methods

2.1. Sampling

The mining sites under study were: Bălan, Deva deposit 1, Deva deposit 2, Brad Ribita and Brad Criscior. For each mining site homogeneous and representative samples were prepared for subsequent characterization. To this end, after being dried in oven at 80 °C for one day and sieved to 4 mm, the fractions were then homogeneously and representatively sampled using a RETSCH rotary splitter. From each site, eight samples were collected.

For the characterization of the particles sizes, the prepared samples were quartered with a manual sampler and submitted to wet sieving using sieves of following sizes: 0.5 mm, 0.351mm, 0.250 mm, and 0.125 mm. The obtained granulometric fractions were filtered, dried in a laboratory oven at 80 °C for one day and then weighed. On the basis of the obtained weights, the distribution curves were constructed (data not showed here).

For each mineral deposit, one of the eight prepared samples was submitted to gravimetric separation by a flow table. The four fractions obtained (light, intermediate, mixed and heavy) were filtered, dried in a stove at 80 °C for a day and then weighed.

This rotary splitter allows the separation of sample into various fractions whose composition corresponds exactly to that of the initial sample, because only a representative sample of the initial rate can provide significant analytical results.

This procedure ensures a high degree of accuracy and reproducibility. It is used in combination with the vibrating feeder RETSCH DR 100, utilized for the homogeneous and uniform assay during the conveying of the material; this is an automatic process of sampling, without interruptions and loss of material. The speed was monitored and kept constant.

A planetary ball mill agate mod. FRITSCH pulverisette, was used for the fine grinding of the samples. The jars and grinding balls were made of agate to prevent samples contamination.

2.2. Characterization

The mineralogical characterization was carried out by the technique of X-ray diffraction (X-ray diffractometer Bruker, mod. D8 Advance).

The analytical determination of metals and gold content of the fractions obtained from the table, was carried out by Perkin Elmer, mod. 400 optical plasma spectrometer (ICP-AES) with data station. This analysis was performed on the solutions achieved after sample chemical dissolution.

Homogeneous and representative samples of approximately 10 g, submitted to chemical attack, were prepared by rotary and manual splitters and subsequently milled in a planetary mill with agate jars and balls. The milling was performed till the particles sizes was less than 80 μm, this being appropriate for the subsequent leaching tests [11]. The sample for gold recovery tests was chosen considering the gold content.

The experimental work was carried out on the heavy fraction constituted by a mix between Brad Ribita and Brad Criscior samples, which had an average gold content of 3 g/t.

2.3. Physical Process

The samples from mining sites Brad Ribita and Brad Criscior, chosen for their higher gold content, were sieved to −0.5 mm, and the fractions submitted to gravimetric separation by the flow table. The table was set with the aim to obtain a richer heavy fraction The heavy fractions recovered were filtered and dried in a stove at 80 °C for one day.

2.4. Grinding

The heavy fractions, obtained by gravimetric separation, were subjected to comminution (<80 μm), using a bar mill, to make them suitable to determination of the content of gold and for the subsequent leaching tests. The grain sizes obtained were analyzed with the SYMPATEC laser granulometer.

At the end of the comminution, the drum was emptied and the slurry after being filtered was dried in a stove at 80 °C. The heavy fractions, after comminution, were mixed homogeneously using the rotary splitter to prepare samples of the mix Brad Ribita-Brad Criscior, and submitted to leaching tests. The gold content of the mixture of minerals was determined after chemical dissolution, with an Atomic Absorption Spectrometer (AAS Perkin Elmer mod. 460).

2.5. Chemical Process

The study of the thiosulphate process, was conducted in mechanical stirred reactors made of Pyrex glass that have a capacity of 2000 mL. Leaching experiments were carried out to study the influence of the concentration of ammonia and the concentration of thiosulphate on gold dissolution, using reagents of analytical grade and distilled water.

The leaching solutions consist of sodium thiosulfate ($Na_2S_2O_3 \cdot 5H_2O$), used as active leaching agent, ammonia (NH_4OH 30%)—for the control of pH—and copper (II) sulphate ($CuSO_4 \cdot 5H_2O$)—which acts as an oxidant of gold [11,16,17].

The tests were carried out at atmospheric pressure and room temperature, while the speed of mechanical agitation was kept constant at 400 rev/min. The leaching time was 4 h, the weight of the samples of 500 g, the particle size minus 80 μm, pH of 10.5 and a redox potential +0.1 V. At set time intervals, small volumes (10 mL) of leaching solution were taken from the reactor. These were analyzed to determine their Au content and, therefore for the kinetic study of gold dissolution [19,20].

The pH and the oxidation-reduction potential of the slurry, were measured using a combined glass electrode and a platinum combination electrode, respectively, both being connected to a digital pH meter.

At the end of each test, the reactor was emptied, while the filtration of the slurry was realized through pressure filters. The solid residue was submitted to washing with distilled water and ammonia;

moreover, gold content was determined, after chemical dissolution of homogeneous and representative samples, with an Atomic Absorption Spectrometer (AAS Perkin Elmer mod. 460).

After leaching, the gold was purified by selective adsorption onto granular activated coconut carbon. The influence of the carbon concentration was studied.

This was conducted in Pyrex glass reactors of capacity of 2000 mL under mechanical stirring (400 rev/min), at room temperature, for a total contact time of 1 hour [20–22].

The tests were carried out by placing in contact leached solution (500 mL) with the different amounts of coconut charcoal, an activated carbon material. The influence of the contact time was investigated by performing liquor withdrawals at set intervals. The concentration of the activated carbon was varied from 5 g/L to 15 g/L. After each experiment, carbon was recovered from the solution and left air drying. Representative samples of carbon were collected and submitted to quantitative chemical analysis.

The desorption of gold from carbon was carried out by elution with a water-ethyl alcohol solution prepared using absolute ethanol (C_2H_5OH) [22]. The gold stripping tests were conducted in a Pyrex glass reactor, with a capacity of 250 mL. The reactor was fitted with three necks: the first one had reflux condenser for the removal of vapors, the second one housed a probe that was connected to a shaking-heating plate for stabilizing the temperature and a thermometer for temperature control was inserted through the third port. The tests were conducted varying the temperature from 40 to 85 °C.

Sampling was performed also at determined time intervals and then chemically analyzed to determine their Au content. These data were used to determine the kinetics of the process.

The final recovery of purified metallic gold from the water-alcohol solution, was carried out by the electrochemical process [19,20,23].

Gold recovery conducted in an electrolytic cell with a capacity of 200 ml in a jacketed Pyrex glass, connected to a Julabo, mod. 5B thermostat (control from −20 to +100 °C). The cell was fitted with a saturated calomel reference electrode, a working electrode (cathode) constituted by a net of platinum wire, having a surface area of 100 cm^2, and a counter-electrode (anode) consisting of a spiral platinum.

The cell was connected to a AMEL, model 555 B potentiostat-galvanostat. The current flowing through the cell was converted into a numerical value by an AMEL, model 721 integrator. The potential difference between the cathode and the anode was measured with an AMEL, model 631 differential electrometer.

The presence of a magnetic stirrer bar allows stirring of the solution into the cell. The electrolysis tests were conducted using 200 mL of strip solution.

3. Results

3.1. Physical Process

The homogeneous and representative samples, achieved after screening, to retain particles with a diameter greater than 0.5 mm, were submitted to gravimetric separation by the flow table. The goal of the physical process was to concentrate pyrite, and then the gold associated with it, in the heavy fraction. The light fraction consists predominantly of quartz.

The heavy fraction for none of the deposits reaches 10%, therefore we must consider that most of the pyrite could be concentrated in the mixed fraction. For all sites the light fraction exceeds 40%, with over 61% in Balan samples.

The chemical analysis, allowed the gold content of the mixed fractions and heavy minerals to be studied. Based on this data the mining sites of greatest interest from the point of view of the gold content are: Brad Ribita and Brad Criscior. These sites were chosen to the study the process of Au recovery. The X-ray diffraction performed on the heavy and mixed fractions from Brad Ribita and Brad Criscior allowed the mineralogical composition to be determined. In detail, the mineralogical species contained in the heavy fraction Brad Ribita are the following: quartz—(SiO_2) (48.3%); pyrite (FeS_2) (20.4%); muscovite—$(K, Ba, Na)_{0.75}(Al, Mg, Cr, V)_2(Si, Al, V)_4O_{10}(OH, O)_2$ (15.5%);

albite—$(Na_{0.75}Ca_{0.25})(Al_{1.26}Si_{2.74}O_8)$ (8.7%); chamosite—$(Mg_{5.036}Fe_{4.964})Al_{2.724}(Si_{5.70}Al_{2.30}O_{20})(OH)_{16}$ (4.9%); Calcite—$CaCO_3$ (2.1%) and Chalcopyrite—$CuFeS_2$ (0.1%).

The following main elements have been determined by chemical analysis of the heavy fraction Brad Ribita: Si (27.0%), S (11.0%), Fe (10.6%), Al (3.3%), V (3.1%), Ca (1.2%), Ba (1.2%), Mg (0.9%), Na (0.8%).

Regarding heavy Brad Criscior, the mineralogical composition is constituted by: quartz—(SiO_2) (66.2%); pyrite (FeS_2) (10.2%); muscovite—$(K, Ba, Na)_{0.75}(Al, Mg, Cr, V)_2(Si, Al, V)_4O_{10}(OH, O)_2$ (14.2%); albite—$(Na_{0.75}Ca_{0.25})(Al_{1.26}Si_{2.74}O_8)$ (5.1%); chamosite—$(Mg_{5.036}Fe_{4.964})Al_{2.724}(Si_{5.70}Al_{2.30}O_{20})(OH)_{16}$ (2.9%); Calcite—$CaCO_3$ (1.1%) and Chalcopyrite—$CuFeS_2$ (0.3%).

Main elements detected by chemical analysis of the heavy fraction Brad Criscior are: Si (33.9%), S (5.5%), Fe (5.4%), V (2.9%), Al (2.5%), Ba (1.1%), Cr (0.8%), Ca (0.6%), Mg (0.7%), Na (0.5%), K (0.3%), Cu (0.1%).

FeS_2 in the mixed fraction for Brad Ribita is 6.3%.

The results of the gravimetric separation are shown in the following Table 1.

Table 1. Heavy fractions separated from the flow table.

Fraction	%
Heavy Brad Ribita	8.0
Heavy Brad Criscior	12.0

3.2. Chemical Process

3.2.1. Gold Leaching

The experimental work was carried out on the heavy fraction, mix of the Brad Criscior and Brad Ribita samples, with an average gold content of 3 g/t. Gold was recovered by leaching with sodium thiosulphate as previously described.

After the preliminary experiment—carried out using conditions from the literature [11,15]—leaching enabled a gold recovery of 38.10% (for the calculation of efficiency there was considered also the gold content within the washing solution and solid residue). The investigation permits the influence and the interactions between $S_2O_3^{2-}$, $CuSO_4$ and NH_3 concentration on gold kinetics to be studied (Figures 2 and 3).

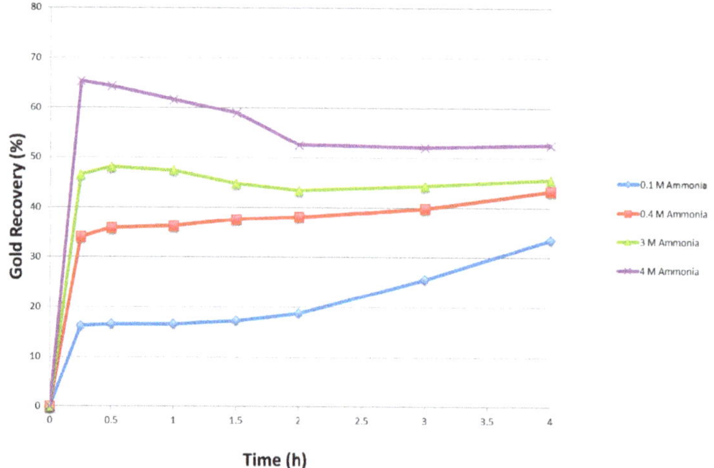

Figure 2. The influence of NH_3 concentration on gold recovery (2 M $Na_2S_2O_3$; 0.1 M $CuSO_4$).

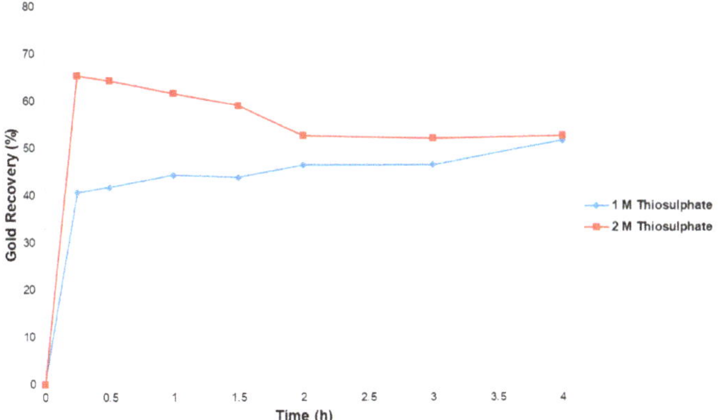

Figure 3. The influence of $S_2O_3{}^{2-}$ concentration on gold recovery (4 M NH_3; 0.1 M $CuSO_4$).

Figure 2 shows the influence of NH_3 concentration on gold recovery, while Figure 3 shows the influence of $S_2O_3{}^{2-}$ concentration. In both cases, $CuSO_4$ concentration was constant at 0.1 M [11,14,22]: the thermodynamic condition that allows the best gold recovery was achieved with a solution of the following composition: 2 M $S_2O_3{}^{2-}$, 0.1 M $CuSO_4$ and 4 M NH_3.

Table 2 describes the best kinetic with higher gold recoveries (about 75% Au after 15 min, without washing). It was observed that the kinetics of extraction decreases by approximately 20% at the end of the experiment (about 65% Au after 4 h, with washing) [11].

Table 2. Gold recovery obtained with a ammoniacal leaching solution having the following composition: 2 M $S_2O_3{}^{2-}$, 0.1 M $CuSO_4$ and 4 M NH_3.

Time (h)	Au (mg)	Au Recovery (%)
0.25	0.91	73.38
0.5	0.90	72.31
1	0.86	69.63
1.5	0.82	66.95
2	0.74	60.52
3	0.73	59.99
4	0.74	60.52
Washing	0.033	2.33

The following Table 3 shows the parameters that allowed the best gold leaching kinetics.

Table 3. Best process parameters of the gold leaching.

Parameters	Values
Temperature (°C)	25
Particle size (μm)	<80
Pulp density (w/v %)	50
pH	10.5
Stirring conditions (rpm)	400
Time of treatment (h)	<1
$S_2O_3{}^{2-}$	2 M
$CuSO_4$	0.1 M
NH_3	4 M

The goals of the further work will be to optimize the leaching process conditions, with the aim to increase gold extraction yields, and to reduce the reagents consumption, such as the concentration of thiosulphate used.

3.2.2. Gold Adsorption

The leaching solution containing gold in the form of soluble complex, is placed in contact with the activated carbon to selectively separate the gold by adsorption. The experimental conditions described in Section 2, were applied to study the influence of the mass ratio of carbon to solution on the gold recovery process. The purification process of the solutions achieved after leaching allowed high gold recoveries to be obtained (Table 4).

Table 4. Kinetics of Au adsorption at different concentrations of carbon in solution.

Time (min)	5 g/L Carbon % Au	10 g/L Carbon % Au	15 g/L Carbon % Au
15	53.59	50.00	82.05
30	61.54	70.52	98.72
45	76.65	87.18	98.72
60	85.90	98.72	98.72

The results show an almost complete recovery of the gold present in solution. From the trend it is clear that the increase of the concentration of carbon in solution favors the recovery. In particular, after 1 h, at a concentration of carbon of 5 g/L, about 86% Au was adsorbed, but the recoveries reach 99% Au when the concentration of the adsorbent increases to 10 g/L. Also it was found with a concentration of 15 g/L, that 99% Au adsorption was achieved after only 30 min.

The parameters that allowed the best gold adsorption onto activated carbon have been reported in Table 5.

Table 5. Best process parameters of gold adsorption on activated carbon.

Parameters	Value
Temperature (°C)	25
Carbon concentration (g/L)	15
pH	10.5
Stirring conditions (rpm)	400
Time of treatment (min)	30

3.2.3. Gold Desorption

The purpose of desorption was to re-extract the gold adsorbed and concentrate it. The duration of process was fixed at 6 h.

From the experimental results, shown in Table 6, it can be observed that the final gold recovery was of 79.0%: further studies using other types of alcohol such as isopropyl alcohol and ethylene glycol, may improve the efficiency, shortening the duration of the stripping process.

Table 6. Kinetics of Au desorption at different concentrations of carbon in solution.

Time (h)	Au (mg)	Au Yield (%)
1	0.20	18.0
2	0.34	30.0
4	0.46	41.0
6	0.88	79.0
Washing	0.30	20.0

The process parameters for the desorption of gold from activated carbon that permitted to achieve 99.0% Au recovery—including washing—after 6 h, are shown in Table 7.

Table 7. Best process parameters of gold desorption from the activated carbon.

Parameters	Value
Temperature	80 °C
Speed of agitation	300 rpm
Time of extraction	6 h
Mass ratio of coal and stripping solution	75 g/L

3.2.4. Electrowinning

The last step of the process was electrowinning. The goal of this step was the recovery of gold from the strip solution by cathodic deposition of the metal. Table 8 shows the experimental results obtained. Table 9 shows the optimized process parameters of the phase of gold electrodeposition.

Table 8. Experimental gold electrowinning parameters.

Time (min)	Cathodic Voltage (V)	Cell Voltage (V)	Current Intensity (mA)	Electric Charge (C)
15	1.4	2.25	100	161
30	1.4	2.51	210	254
45	1.4	2.47	210	352
60	1.4	2.52	210	448
75	1.4	2.51	210	564

Table 9. Best process parameters of the phase of gold electrodeposition.

Parameters	Value
Temperature	40 °C
Time of electrolysis	75 min
Voltage at the cathode	−1.4 V
Voltage of the cell	2.50 V

The kinetics of electrodeposition in cell of laboratory is quick and the final metallic gold recovery was high (98% Au). The concentration of gold in the sterile solution was below the detection limit of the instrument) after the first 30 min. A dark deposit was uniformly distributed on the surface of the cathode.

From the reported data, it was noted that the intensity of the measured current is around 210 mA, which corresponds to a current density of about 2.1 mA/cm^2, given that the cathode surface is 100 cm^2. As can be seen from the experimental results, the amount of charge that passes through the cell after 30 min was 254 Coulombs. Hence the current efficiency was low (5%); this is due to parasitic reactions, such as the reduction of the water and of dissolved oxygen. The consumption of energy was high; about 20 kWh/kg of gold deposited.

4. Discussion

The preliminary study of the parameters and operating conditions for the various stages of the thiosulphate process shows, for the sample composed of mining wastes from Brad Ribita and Brad Criscior, the technical viability of the process. The experimental results obtained indicate good gold dissolution kinetics in the aqueous ammoniacal solution of thiosulphate, which can be used without special precautions and restrictions.

The gold extraction reached an average final value of 75% Au, working at room temperature, but, the trends for dissolution demonstrate that the thermodynamic parameters were not optimized, because gold recovery decreases during the best experiment. This fact, probably, is due to the thermodynamic instability of the complexing ligands (ammonia and thiosulphate) and oxidizing agents (copper ions) present in the system, tested for the first time on this type of material [11,24].

It is noted that the extraction of gold is already very high in the first 15 minutes: this is due to the gravimetric enrichment and means that a part of the gold is free and is not incorporated into the mineral matrix (see Table 2).

Gold recoveries for the overall process including leaching (extraction yield of about 75% Au) and complete adsorption-desorption–electrodeposition cycle (about 90% Au recovered) were about 65–67% (considering the gold content in the washing) in line with the conventional cyanidation process, as demonstrated in previous experimental work [25].

These results are very encouraging, considering that it is commercially an innovative process, applied to a gold ore with a low content.

Samples were leached after comminution <80 μm: a finer grinding would probably result in an increase in gold extraction, but its convenience can only be determined after careful cost analysis.

The optimization of the process is still required to identify the best process parameters and operating conditions. Considering the progressive depletion of gold mining reserves and the inability of the gold production to quickly react to the prospect of a change in prices and to changes in demand, it is providential to recover gold from mining tailings. The high price of the precious metal allows, as preliminary economic analysis shows, the feasibility of alternative processes despite the low levels of gold, the large amount of sterile to be treated and the high costs of extraction.

The enhancement of the tailings, in addition to the sale of raw materials of high grade, allows to implement an effective and sustainable recovery technology. In this way it is possible to guarantee over time the use of two indispensable resources of primary importance: the environment as a whole on one side and raw materials mining the other.

In addition to the economic and environmental aspects, we must consider the social benefits that the application of these innovative processes may generate such as providing many jobs and contributing thus to the development of repressed areas, improving the competitiveness of and creating added value and new jobs in raw materials processing, refining, equipment manufacturing and downstream industries.

In the next step of investigation, pre-treatment of components such as pyrite will be investigated before the extraction of gold. For this purpose a treatment circuit that employs biotechnological processes, with low energy consumption, will be integrated.

The elimination of pyrite will help to reduce the cost of extracting gold, reducing the consumption of the reagents. The application of physical-chemical and biological-chemical methods will allow the treatment of the Acid Mine Drainage, together with the recovery of heavy metals such as copper [26]; moreover, the optimization of leaching kinetics will be performed. The complete leaching process analysis will be outlined, including detailed description of the process scheme together with the economic analysis.

5. Conclusions

The present work examined a potential innovative application of a treatment for gold recovery, based on the thiosulphate process.

The challenge here was to apply it to the recovery from resources with low gold content. The preliminary application of this circuit allowed the following results in the various steps to be obtained:

(1) Leaching: the gold extraction reached a final value of about 75% Au at room temperature.
(2) Purification (adsorption/desorption on/by activated carbon and electrowinning) about 90% Au.
(3) Recoveries of the overall process of about 65–67% Au, in line with the conventional cyanidation process.

These results are very encouraging, considering that this is a commercially innovative process, applied to a low gold content ore.

The next objective is to study parameters that allow the improvement of the gold dissolution kinetics and the subsequent steps of recovery from purified solutions, thereby determining beforehand the technical feasibility of the scheme of process developed at the laboratory scale.

The optimization of process parameters and operating conditions, and scale up of the process at industrial level will permit the best results in terms of process yields to be achieved, and in turn will allow us to exploit important resources for the European economy.

Author Contributions: Conceptualization, F.V. and V.G.; Data curation, F.V. and D.G.; Formal analysis, S.U. and F.V.; Investigation, D.G.; Methodology, D.G.; Supervision, S.U.; Writing—original draft, S.U.; Writing—review & editing, S.U. and V.G.

Funding: The work was supported financially by the Istituto di Geologia Ambientale e Geoingegneria, CNR, through special funds for free theme research.

Acknowledgments: This experimental work was carried out during the course of the experimental degree thesis of Dr. Alessia Panone from Università degli Studi dell'Aquila. The authors are grateful to Dr. Alessia Panone and Mr. Pietro Fornari for their helpful collaboration during the experimental work.

Conflicts of Interest: The authors declare no conflict of interest.

References

1. Obiri, S. Determination of heavy metal in water from boreholes in Dumasi in the Wassa West District ofWestern Region of Republic of Ghana. *Environ. Monit. Assess.* **2007**, *130*, 455–463. [CrossRef] [PubMed]
2. Saldarriaga-Isaza, A.; Villegas-Palacio, C.; Arango, S. The public good dilemma of a non-renewable common resource: A look at the facts of artisanal gold mining. *Res. Policy* **2013**, *38*, 224–232. [CrossRef]
3. Ako, T.A.; Onoduku, U.S.; Oke, S.A.; Adamu, I.A.; Ali, S.E.; Mamodu, A.; Ibrahim, A.T. Environmental impact of artisanal gold mining in Luku, Minna, Niger State, North Central Nigeria. *J. Geoscienc. Geomatics* **2014**, *2*, 28–37.
4. Bawua, S.A.; Owusu, R. Analyzing the effect of Akoben Programme on the environmental performance of mining in Ghana: A case study of a Gold Mining Company. *J. Sustain. Min.* **2018**, (in press). [CrossRef]
5. Piga, L.; Abbruzzese, C.; Fornari, P.; Massidda, R.; Ubaldini, S. Thiourea leaching of a siliceous Au-Ag bearing ore using a four-factor composite design. In Proceedings of the XIX International Mineral Processing Congress, San Francisco, CA, USA, 22–27 October 1995; Society for Mining, Metallurgy, and Exploration: Littleton, CO, USA, 1995.
6. De Michelis, I.; Olivieria, A.; Ubaldini, S.; Ferella, F.; Beolchini, F.; Vegliò, F. Roasting and chlorine leaching of gold-bearing refractory concentrate: Experimental and process analysis. *Int. J. Min. Sci. Technol.* **2013**, *23*, 709–715. [CrossRef]
7. *Annual Report 2011: Geological Romanian Institute;* Geological Romanian Institute—National Geological Survey of Romania: Bucharest, Romania, 2011.
8. *Annual Report 2011;* Romanian National Agency Mineral Resources (ANRM): Bucharest, Romania, 2011.
9. Ubaldini, S.; Vegliò, F.; Toro, L.; Abbruzzese, C. Gold recovery from Pyrrhotite by Bioleaching and Cyanidation: A preliminary study using statistical methods. *Miner. Bioprocess.* **1995**, *2*, 145–155.
10. Ubaldini, S.; Massidda, R.; Abbruzzese, C.; Vegliό, F.; Toro, L. Gold recovery from finely disseminated ore by use of cyanidation and thioureation. In Proceedings of the 6th International Mineral Processing Symposium, Kuşadası, Turkey, 24–26 September 1996; Balkema: Rotterdam, The Netherlands, 1996.
11. Abbruzzese, C.; Fornari, P.; Massidda, R.; Vegliò, F.; Ubaldini, S. Thiosulphate leaching for gold hydrometallurgy. *Hydrometallurgy* **1995**, *39*, 265–276. [CrossRef]
12. Aylmore, M.G.; Muir, D.M. Thiosolfate leaching of gold—A review. *Miner. Eng.* **2001**, *14*, 135–174. [CrossRef]
13. Grosse, A.C.; Dicinoski, G.W.; Shaw, M.J.; Haddad, P.R. Leaching and recovery of gold using ammoniacal thiosulfate leach liquors. *Hydrometallurgy* **2003**, *69*, 1–21. [CrossRef]
14. Rath, R.; Hiroyosh, N.; Tsunekawa, M.; Hirajima, T. Ammoniacal thiosulphate leaching of gold ore. *Eur. J. Miner. Process. Environ. Prot.* **2003**, *3*, 344–352.

15. Ubaldini, S.; Veglio', F.; Massidda, R.; Abbruzzese, C. A new technology for gold extraction from activated carbon after cyanidation. In Proceedings of the XXII International Mineral Processing Congress (IMPC), Cape Town, South Africa, 29 September–3 October 2003; South African Institute of Mining & Metallurgy: Johannesburg, South Africa, 2003.
16. Bas, A.D.; Ozdemir, E.; Yazici, E.Y.; Celep, O.; Deveci, H. Ammoniacal thiosulphate leaching of a copper-rich gold ore. In Proceedings of the 15th Conference on Environment and Mineral Processing, VŠB-TU Ostrava, Czech, 8–10 June 2011.
17. Liu, X.; Xu, B.; Yang, Y.; Li, Q.; Jiang, T.; Zhang, X.; Zhang, Y. Effect of galena on thiosulfate leaching of gold. *Hydromettallurgy* **2017**, *171*, 157–164. [CrossRef]
18. Seisko, S.; Lampinen, M.; Aromaa, J.; Laari, A.; Koiranen, T.; Lundstrom, M. Kinetics and mechanisms of gold dissolution by ferric chloride leaching. *Miner. Eng.* **2018**, *115*, 131–141. [CrossRef]
19. Ubaldini, S.; Fornari, P.; Massidda, R.; Abbruzzese, C. An innovative thiourea gold leaching process. *Hydrometallurgy* **1998**, *48*, 113–124. [CrossRef]
20. Ubaldini, S.; Massidda, R.; Abbruzzese, C.; Veglio', F. A cheap process for gold recovery from leached solutions. *Chem. Eng. T.* **2003**, *3*, 485–490.
21. Jeffrey, M.I.; Hewitt, D.M.; Dai, X.; Brunt, S.D. Ion exchange adsorption and elution for recovering gold thiosulfate from leach solutions. *Hydrometallurgy* **2010**, *100*, 136–143. [CrossRef]
22. Ubaldini, S.; Massidda, R.; Veglio', F.; Beolchini, F. Gold stripping by hydro-alcoholic solutions from activated carbon: Experimental results and data analysis by a semi-empirical model. *Hydrometallurgy* **2006**, *81*, 40–44. [CrossRef]
23. Ubaldini, S.; Veglio', F.; Toro, L.; Abbruzzese, C. Combined bio-hydrometallurgical process for gold recovery from refractory stibnite. *Miner. Eng.* **2000**, *13*, 1641–1646. [CrossRef]
24. Povar, I.; Ubaldini, S.; Lupascu, T.; Spinu, O.; Pintilie, B. *The Solution Chemistry of the Copper(II)-Ammonia Thiosulfate Aqueous System. Proceedings Book of the 21st International Symposium "The Environment and the Industry", SIMI 2018, Bucharest, Romania*; National Research and Development Institute for Industrial Ecology–ECOIND: Bucharest, Romania, 2018.
25. Panone, A. Valorisation of mining landfills in Romania: Process of gold recovery from tailings. Master's Thesis, Faculty of Engineering of the University of L'Aquila, L'Aquila, Italy, 2011.
26. Luptakova, A.; Ubaldini, S.; Macingova, E.; Fornari, P.; Giuliano, V. Application of Physical-chemical and Biological-chemical Methods for Heavy Metals Removal from Acid Mine Drainage. *Process Biochem.* **2012**, *47*, 1633–1639. [CrossRef]

Article

Leaching Chalcopyrite Concentrate with Oxygen and Sulfuric Acid Using a Low-Pressure Reactor

Josué Cháidez [1,*], José Parga [2], Jesús Valenzuela [3], Raúl Carrillo [4] and Isaías Almaguer [5]

1. Metallurgical Processes, Servicios Especializados Peñoles S.A de C.V., Torreón 27300, México
2. Division of Graduate Studies and Research, Instituto Tecnológico de Saltillo, Saltillo 25280, México; jrparga@itsaltillo.edu.mx
3. Department of Chemical and Metallurgical Engineering, Universidad de Sonora, Hermosillo 83000, México; jvalen@iq.uson.mx
4. Faculty of Metallurgy, Universidad Autónoma de Coahuila, Monclova 25710, México; frrcarrillo@yahoo.com.mx
5. Metallurgical Processes, Servicios Administrativos Peñoles S.A. de C.V., Torreón 27300, México; isaias_almaguer@penoles.com.mx

* Correspondence: josue_chaidez@hotmail.com; Tel.: +52-871-121-4378

Received: 7 January 2019; Accepted: 1 February 2019; Published: 6 February 2019

Abstract: This article presents a copper leaching process from chalcopyrite concentrates using a low-pressure reactor. The experiments were carried out in a 30 L batch reactor at an oxygen pressure of 1 kg/cm^2 and solid concentration of 100 g/L. The temperature, particle size and initial acid concentration were varied based on a Taguchi L9 experimental design. The initial and final samples of the study were characterized by chemical analysis, X-ray diffraction and particle size distribution. The mass balance showed that 98% of copper was extracted from the chalcopyrite concentrate in 3 h under the following experimental conditions: 130 g/L of initial sulfuric acid concentration, temperature of 100 °C, oxygen pressure of 1 kg/cm^2, solid concentration of 100 g/L and particle size of −105 + 75 μm. The ANOVA demonstrated that temperature had the greatest influence on copper extraction. The activation energy was 61.93 kJ/mol. The best fit to a linear correlation was the chemical reaction equation that controls the kinetics for the leaching copper from chalcopyrite. The images obtained by SEM showed evidence of shrinking in the core model with the formation of a porous elemental sulfur product layer.

Keywords: Hydrometallurgical processes; Chalcopyrite; kinetics; low-pressure leaching

---

1. Introduction

Chalcopyrite is the most abundant sulfide copper mineral in the Earth's crust. Generally, it is associated with other compounds such as galena, sphalerite, pyrite, arsenic, antimony or bismuth sulfides; moreover, it is often bonded with valuable metals such as silver and gold.

From an environmental and economic perspective, further technological developments for obtaining high-grade copper in an efficient and cost-effective manner are desirable. Today, companies such as Beijing Nonferrous Metal, JX Nippon Mining & Metals, Freeport McMoran, Freeport Minerals, Phelps Dodge, Outotec, BHP Billiton, etc. are investing in hydrometallurgical research because of the potential associated economic benefits [1].

Specifically, hydrometallurgical processes have a series of advantages in comparison to pyrometallurgical processes, for example, the required plant capacity is smaller (<10,000 t/y of copper), there is no need for an acid plant, no dust is emitted, etc.

Hydrometallurgical pilot plant projects such as Outotec, Galvanox, Activox and AAC/UBC have implemented some of the latest technology for mineral leaching, and several demo plants have also been installed. In particular, leaching reactors have been developed by the hydrometallurgical industry, wherein an oxidant catalyzer is commonly introduced into a pressure reactor to leach copper under varying temperatures and pressures. Table 1 lists the existing hydrometallurgical processes for leaching copper in a sulfate media, which are classified according to low, medium and high temperatures and pressures [1]. However, other aqueous media have been studied for the chalcopyrite leaching (glycine [2], nitrate [3], chloride [4], ammonium [5], etc . . .).

Most commercial plants operate under conditions of high temperature and pressure in a sulfate medium. Nevertheless, in Las Cruces (Spain), a commercial plant with an atmospheric leaching copper process has been implemented with Outotec technology [6].

The present article focuses on the leaching stage of the hydrometallurgical process to recovery copper and iron in the liquid phase using a batch reactor under low temperature and oxygen pressure conditions. In subsequent processing, lead, silver and gold may be recovered in the resulting residue by the pyrometallurgical process of lead [7–9]. The design of this technology was based on the fundaments of thermodynamics and metallurgy.

Table 1. Current hydrometallurgical processes for leaching copper. Information obtained from [1].

Leaching Processes	Name of the Processes	Country/Company	Status	Copper Mineral	Temperature (°C)	Pressure (atm)	Grinding	Acid	Oxidant Catalizer	Production (t/year)
Sulfate Medium	Mount Gordon	Australia/Aditya Birla	Commercial plant	Chalcocite with pyrites	80–90	8	80%, 100 µm	Diluted H_2SO_4	O_2, ions Fe^{3+}	50,000
	Activox	Botswana (Tati)/Norilsk Process Technology	Pilot plant	Nickel-copper concentrates	90–110	10–12	Ultrafine (5–10 µm)	Diluted H_2SO_4	O_2, ions Fe^{3+}	12,000–16,000
Low temperature and low-medium pressure	Las Cruces	Spain/First Quantum Minerals	Commercial plant	Chalcopyrite	90	Atmospheric	10–15 µm	Diluted H_2SO_4	H^+, O_2 and Fe^{3+}	72,000
	Galvanox	Canada (Vancouver)/UBC	Pilot plant	Chalcopyrite or enargite with pyrite	80	Atmospheric	75 µm	Diluted H_2SO_4	O_2 or air, pyrite or silver	-
Sulfate Medium	Anglo American Corporation/University of British Columbia (AAC/UBC)	South Africa (Johannesburg)/AAC-UBC	Pilot plant	Chalcopyrite	150	10–12	80%, 10 µm	Diluted H_2SO_4	O_2, surfactants (grinding required)	-
Medium temperature and low-medium pressure	Freeport McMoRan	USA (Arizona)/ Freeport McMoRan	Commercial plant	Copper sulphides concentrates	160	13.6	98%, 15 µm	Diluted H_2SO_4	Surfactants and O_2	65,200
Sulfate Medium	Freeport McMoRan	USA (Arizona)/Freeport McMoRan	Semi-commercial plant (now closed)	Chalcopyrite and molybdenite	225	32.5	Fine grinding	Diluted H_2SO_4	O_2	16,000
High temperature and high Pressure	Sepon Copper	Sepon/MMG	Commercial plant	Chalcocite and clays	80	1	100 µm	Diluted H_2SO_4	Sulfuric acid, Fe^{3+} ions	90,000
				Pyrite	230	30–32	80%, 50 µm	Diluted H_2SO_4	O_2	
Bioleaching	BioCop	Chile (Chuquicamata)/Alliance copper (BHP Billiton y CODELCO)	Commercial plant	Chalcopyrite and enargite	70–80	Atmospheric	37 µm	Diluted H_2SO_4	O_2, thermophile extreme bacteria, Fe^{3+} ions	20,000
	BacTech-Mintek	México/Peñoles	Demo plant	Chalcopyrite and copper sulphides	35–50	Atmospheric	10–20 µm	Diluted H_2SO_4	Air, moderate thermophile bacteria ions Fe^{3+}	160

2. Materials and Methods

2.1. Material and Equipment

The experiments were carried out in a stainless steel 316 L closed reactor having a volume of 30 liters and being equipped with an agitation system with 4 baffles, a security valve calibrated at 2 kg/cm^2, a rupture disc calibrated at 3 kg/cm^2, a pressure transmitter with a chemical seal and a resistance temperature detector (RTD) connected to a data logger. Also, the reactor had a controlled cooling-heating jacket. Figure 1 shows an image of the reactor.

Figure 1. Batch reactor for leaching chalcopyrite concentrate.

The chalcopyrite concentrate was supplied by Peñoles (Mexico). Samples of the concentrate were characterized by X-ray diffraction (XRD, Panalytical, Empyrean model), chemical analysis (CA, PerkinElmer 8300, LECO SC230DR) and a backscattered electron (BSE) module in a scanning electron microscope (SEM, FEI, Quanta600 model) for a wider range of the mineralogical species. The chemical analysis is presented in Table 2. The carbonate content was calculated with the difference of total and organic carbon. Table 3 shows the mineralogical reconstruction via XRD and CA expressed in terms of weight percentage (Wt.%). The mineralogy species obtained by BSE-SEM are shown in Table 4 in terms of weight percentage (Wt.%).

Table 2. Chemical analysis of the chalcopyrite concentrate.

Element	Cu	Fe	As	Pb	Ca	Zn	Al	S	Si	CO_3
Wt.%	24.7	26	0.81	6.56	1.25	6.36	0.27	29.9	1.12	1.13

Table 3. Mineralogical reconstruction of the chalcopyrite concentrate.

Compounds		Weight%
Chalcopyrite	$CuFeS_2$	70.7
Galena	PbS	7.8
Sphalerite	ZnS	9.3
Gypsum	$CaSO_4$	4.1
Pyrite	FeS_2	7.8

Then, the chalcopyrite concentrate was fractionated to different sizes in a Tyler RO-TAP® Sieve Shaker using −74, −105 + 74 and −149 + 105 μm. All fractions were characterized, and no significant variation was observed in the mineralogical composition and chemical analysis. The particle size distribution of the residues was measured in a Horiba LA 950 V2, which expressed the results in terms of the equivalent spherical diameter.

Table 4. Results of the analysis of the chalcopyrite concentrate by the SEM-BSE system.

Group	Mineral	Formula	Weight%
Sulfides	Galena	PbS	9.31
	Sphalerite	ZnS	11.09
	Chalcopyrite	$CuFeS_2$	68.65
	Tetrahedrite	$(Cu_{0.8}Fe_{0.1}Zn_{0.1})_{12}(Sb_{0.8}As_{0.2})_4S_{13}$	0.08
	Pyrite	FeS_2	2.26
	Pyrrhotite	FeS	2.00
	Arsenopyrite	$FeAsS$	1.27
Silver species	Native Ag	Ag	0.16
	Freibergite	$(Ag_{0.3}Cu_{0.6}Fe_{0.1})_{12}Sb_4S_{13}$	0.002
	Enargite	Cu_3AsS_4	0.01
Gangues and other oxides species	Andradite	$Ca_3Fe_2Al(SiO_4)_3$	0.41
	Apatite	$Ca_5(PO_4)_3(F, Cl, OH)$	0.004
	Augite	$(Ca,Mg,Fe)_2(Si,Al)2O_6$	0.27
	Biotite	$K(Mg, Fe)_3AlSi_3O_{10}(OH, F)_2$	0.09
	Calcite	$CaCO_3$	1.35
	Chlorite	$(Mg,Fe)_3(Si,Al)_4O_{10}(OH)_2{\cdot}(Mg,Fe)_3(OH)_6$	0.10
	Quartz	SiO_2	0.48
	Diopside	$CaMgSi_2O_6$	0.23
	Grossularite	$Ca_3Al_2Si_3O_{12}$	0.36
	Moonstone	$(Ca_{0.6}Na_{0.4})Si_2AlO_8$	0.18
	Orthoclase	$K(AlSi_3O_8)$	0.48
	Ox_Fe	Fe_xO_y	0.14
	Titanite	$CaTiSiO_5$	0.02
	Others	-	1.06

2.2. Experimental Method

To clarify the effects of particle size, temperature and initial sulfuric acid concentration on copper extraction, a Taguchi 3^3 experimental design was employed. In addition, a tenth test was done using different temperatures to calculate the activation energy. The following experimental conditions were constant: residence time (7 h), oxygen pressure (1 kg/cm^2), solid concentration (100 g/L) and agitation velocity (550 RPM). Table 5 shows the experimental design.

Table 5. Taguchi L9 experimental design.

No. Test	Particle Size (µm)	Initial Acidity (g/L)	Temperature (°C)
1	−74	100	80
2	−74	130	90
3	−74	155	100
4	−105 + 74	100	90
5	−105 + 74	130	100
6	−105 + 74	155	80
7	−149 + 105	100	100
8	−149 + 105	130	80
9	−149 + 105	155	90
10	−149 + 105	130	50

The experimental procedure began with the addition of hot water (80 °C) to the reactor and the initiation of the agitation system, which was set at a low revolution speed; then, the chalcopyrite concentrate was fed into the reactor, followed by sulfuric acid. After the addition of these materials, a 2 min air purge was performed; then, the reactor was closed and pressurized to 1 kg/cm^2 with medicinal oxygen. The agitation velocity was set at 550 RPM, and the data logger was then turned on to start recording data.

To determine the kinetics of the copper leaching process, 100 mL samples were taken from a lateral valve of the reactor at different time intervals during the test.

3. Results

3.1. Thermodynamics

The thermodynamics of copper sulfide leaching are based on the interaction of the elements required to carry out the decomposition of chalcopyrite. The main reactions that govern the leaching of copper concentrates are shown as follows:

$$CuFeS_2 + 2.5O_2 + H_2SO_4 = CuSO_4 + FeSO_4 + H_2O + S^\circ \quad (1)$$

$$CuFeS_2 + 2Fe_2(SO_4)_3 = CuSO_4 + 5FeSO_4 + 2S^\circ \quad (2)$$

$$4FeSO_4 + O_2 + 2H_2SO_4 = 2Fe_2(SO_4)_3 + 2H_2O \quad (3)$$

The Pourbaix diagrams in Figure 2 show that a low pH is required to keep copper and iron in sulfate solution. In addition, to ensure that ferric ions are present in the solution, the oxide potential must be above 0.57 V (SHE). This step enables indirect leaching via reaction 2, wherein the oxidation-reduction cycle of iron facilitates the decomposition of chalcopyrite. The Pourbaix diagrams were calculated with the software HSC 8.0.6 at 95 °C, [Cu] = 0.787 mol/L, [Fe] = 0.895 mol/L and [S] = 1 mol/L [10].

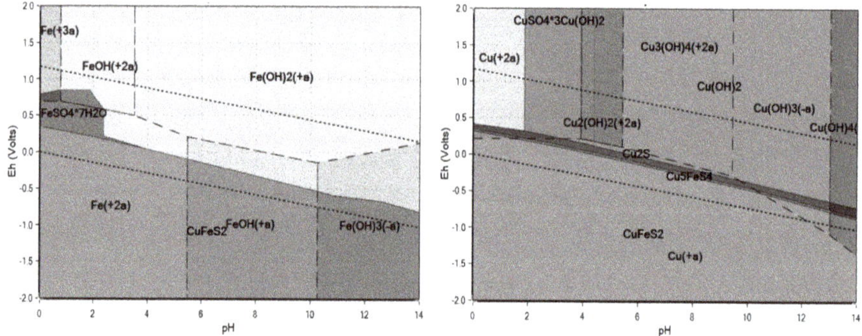

Figure 2. Cu-Fe-S Pourbaix diagrams.

3.2. Extraction and Chemical Analysis

The results of the experiment show that the main influential variables in the leaching process were temperature and initial sulfuric acid concentration. Particle size did not significantly affect the process. The results of copper extraction versus time are presented in Figure 3.

The test number 5 presented the best results according to the mass balance. Under the corresponding conditions, 97.99% of copper was extracted in 3 h of reaction. The solid shrink was 61.8% (Wt.%), and the oxygen consumption was 0.662 g O_2/g Cu fed. The density of the final solution was 1.15 g/mL, and a final oxidation-reduction potential (ORP) of 0.483V was measured in the suspension with a calomel electrode (Hg/Hg_2Cl_2).

Notably, different authors have reported percentages of copper extraction from chalcopyrite of 70% [11], 65% [12], 60% [13], 83% [14] and up to 95% [15]; in addition to 95% via the arbiter process, 98% via the Freeport McMoran method, 97–98% via the Activox process, 98% via the Albion process and 95% via the Galvanox process [1].

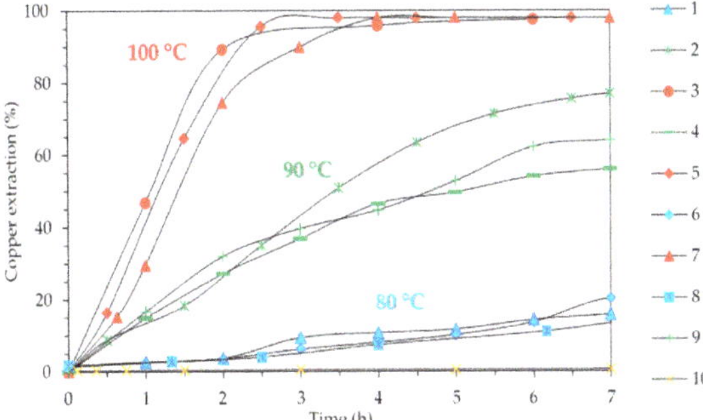

Figure 3. Copper extraction versus leaching time.

The test 5 was carried out at 100 °C with an initial sulfuric acid concentration of 130 g/L; this concentration of sulfuric acid was sufficient for reaction with species and to keep the iron in solution. Table 6 presents the CA of the residues and the solution along with the percentages of elemental distribution in the liquid phase that were obtained from the mass balance.

Table 6. Chemical analysis of the test 5 residue (Wt.%), chemical analysis of the solution (g/L) and the elemental distribution in the liquid phase (%) of the leaching process.

Element	Residue (Wt.%)	Solution (g/L)	Distribution in Liquid Phase (%)
Cu	1.12	22.83	97.99
Fe	3.65	27.26	94.69
As	0.08	0.8	95.9
Pb	13.2	0.055	1.0
Zn	0.25	5.96	98.24
S°	56.2	0.0	0.0
Fe^{2+}	-	3.89	-
H_2SO_4	-	45.45	-

The leaching solution contained a high percentage of zinc, copper and iron because of the solubility of these elements in the utilized sulfate medium (given the temperature and acid concentration). In a global process view, it is important to consider that the solution could be treated in a solvent extraction and electrowinning stages for the production of electrolytic copper. The raffinate from solvent extraction could be neutralized with calcium carbonate to precipitate ferric ions, zinc, arsenic and minor elements; then, the solution could be recycled to the direct leaching stage.

The iron in the residue mainly corresponded with pyrite, which requires higher temperature and pressure for decomposition. Table 7 shows the species present in the solid phase in terms of weight percentages, including elemental sulfur (64.1%), anglesite (19.3%), silica (5%), pyrite (5.7%), unreacted chalcopyrite (3.23%) and gypsum (2.3%).

Table 7. Mineralogical reconstruction of the leaching residue in test 5.

Compounds		Wt.%
Chalcopyrite	$CuFeS_2$	3.23
Anglesite	$PbSO_4$	19.3
Gypsum	$CaSO_4$	2.3
Silica	SiO_2	5.0
Pyrite	FeS_2	5.7
Elemental sulfur	S_8	64.1

From an economic perspective, the recovery of valuable minerals in residues following the hydrometallurgical treatment of chalcopyrite concentrates is important. The high content of elemental sulfur could reduce the profitability of recovering valuable metals by cyanidation or melting processes.

Table 8 presents the final particle size distribution of the P5 test. As expected, the particle size decreased considerably from 100% +74 μm to 90% −16.92 μm because of the leaching of chalcopyrite particles and the formation of elemental sulfur.

Table 8. Particle size distribution of the resulting residue from chalcopyrite leaching in test P5.

Particle Size Distribution		
D90%	D50%	D10%
16.92	11.00	6.99

Figure 4 shows electron images obtained by SEM-BSE of unreacted chalcopyrite and galena particles, which were identified by SEM-EDS. The porous layer of elemental sulfur surrounding the particles can be observed. These particles are indicative of the shrinking core model, wherein a layer of elemental sulfur is formed as a product. However, the high extraction percentage of copper obtained in a short time (3 h) indicates that the elemental sulfur layer does not passivate the leaching of the chalcopyrite concentrate.

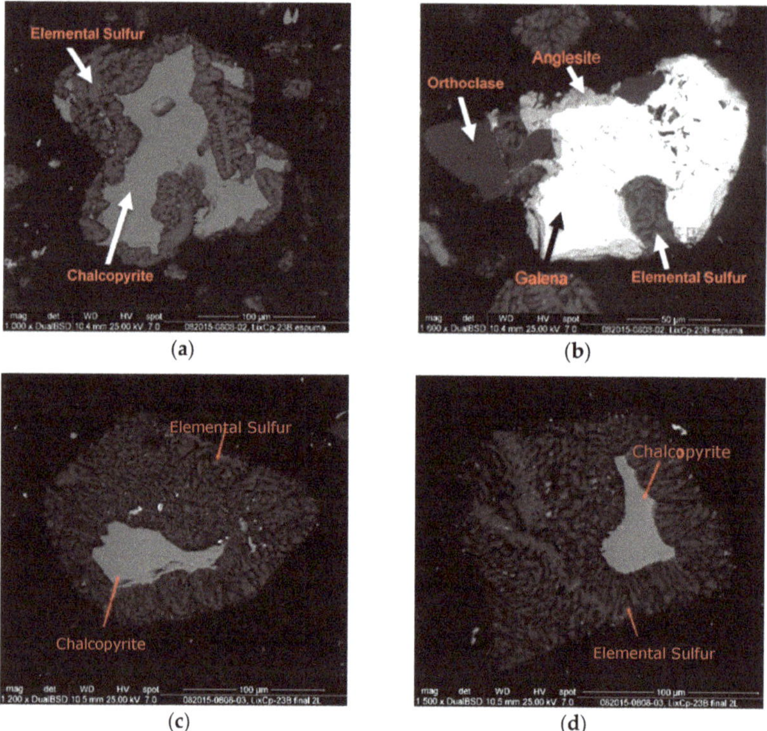

Figure 4. Punctual microanalysis by SEM-BSE-EDS in the leached residue of test P5. (**a**) Unreacted chalcopyrite particle, (**b**) unreacted galena particle, (**c**) unreacted Chalcopyrite particle, (**d**) unreacted Chalcopyrite particle.

Figure 5 shows the microstructure and the X-ray mapping by SEM-EDS for sulfur, copper and iron of the unreacted chalcopyrite particle.

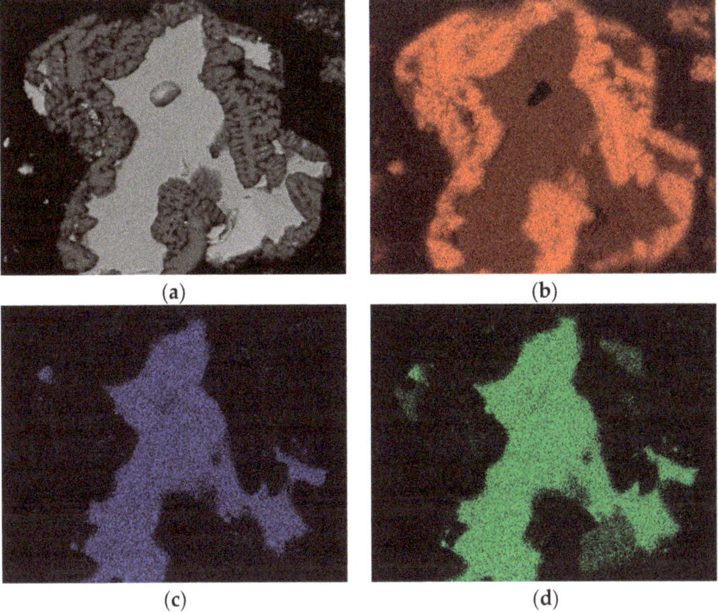

Figure 5. X-ray mapping by SEM-EDS in the unreacted chalcopyrite particle of test P5. (**a**) electron image, (**b**) S, (**c**) Cu, (**d**) Fe.

In Figure 6, the temperature profile, in addition to the partial and accumulated oxygen consumption, are shown. At 2 h, oxygen consumption reaches its maximum and then decreases after this point.

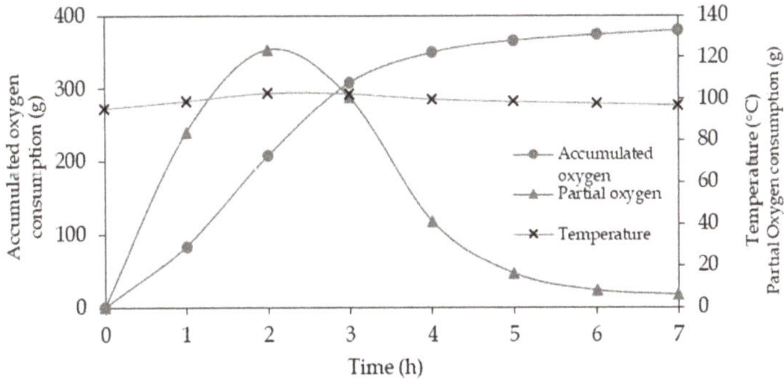

Figure 6. Temperature and partial and accumulative oxygen consumption over time in test P5.

A similar finding was observed for the iron concentration in solution; ferrous ions reached their maximum concentration at 2 h and then decreased. This could explain the chalcopyrite leaching as a two steps process.

Specifically, the first step occurred within the initial 2 h of the test, wherein most of the chalcopyrite was decomposed at temperatures above 95 °C. The second step occurred when the remaining ferrous ions were oxidized to ferric iron. Depending on the next stages, which are related with the liquid phase (iron purification or solvent extraction), the Fe^{3+}/Fe^{2+} relation must be as high as possible to ensure that the process is not affected. In Figure 7, the concentration profile of iron and ferrous ions during leaching is shown.

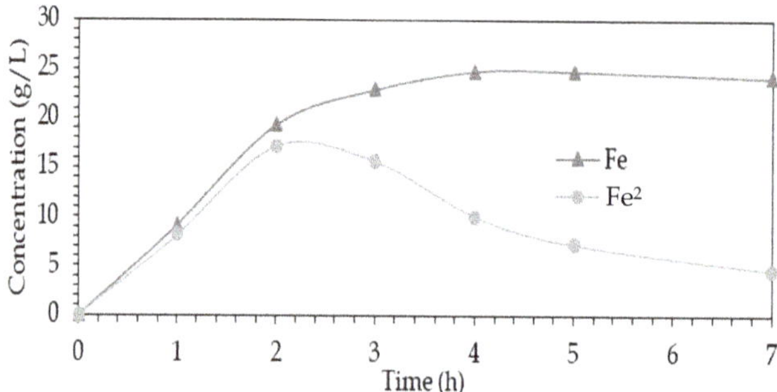

Figure 7. Concentration of iron and ferrous ions over time in test P5.

3.3. Statistical Analysis

As mentioned in Table 5, this study is based on a modified Taguchi L9 experimental design with three levels for three independent variables or parameters. An additional experiment was realized (Test 10): analysis of variance (ANOVA) of the experimental tests data at different conditions was used to evaluate the effect of each individual variable. The results of Test 10 were not included in ANOVA due the difference of temperature between the lowest and the average and the small amount of Cu extraction observed.

Table 9 shows the effects of each parameter using the ANOVA module of the Minitab 15 software. The table shows the values of degree freedom (DF), sum of squares (SS), media of squares (MS), Fisher ratio (F), probability level (Prob Level) and the probability that a false null hypothesis can be rejected (Power) with a 95% confidence level ($\alpha = 0.05$). According to F, Prob Level and Power values, ANOVA shows that under the studied conditions, temperature is the most important factor for copper extraction and oxygen consumption. The results also indicated that within the analyzed range, the other two variables studied (initial acid and particle size) did not have a statistically-significant effect.

Table 9. Analysis of variance (ANOVA) for Cu extraction.

Parameter	DF	SS	MS	F	Prob Level	Power
				Cu Extraction		
Temperature	2	10160.97	5080.487	98.26	0.010075 *	0.993017
Initial acid	2	114.2358	57.11791	1.1	0.47513	0.10105
Particle size	2	36.89769	18.44884	0.36	0.737023	0.066798
S	2	103.4102	51.70508			
Total (Adjusted)	8	10415.52				
Total	9					
				Oxygen Consumption		
Temperature	2	0.2263376	0.1131688	46.99	0.020836 *	0.909376
Initial acid	2	4.52×10^4	2.26×10^4	0.09	0.91428	0.054443
Particle size	2	1.85×10^3	9.25×10^4	0.38	0.722385	0.06808
S	2	4.82×10^3	2.41×10^3			
Total (Adjusted)	8	0.2334562				
Total	9					

* $\alpha = 0.05$.

Figure 8 shows the correlations of temperature, initial acid concentration and particle size with percentage of copper extraction. As observed, temperature had the greatest effect on the process of leaching copper from chalcopyrite concentrates. Figure 9 shows the correlations of temperature, initial acid concentration and particle size with oxygen consumption. As expected, temperature once again had the greatest effect on oxygen consumption, whereas particle size and initial acid concentration had no clear influence.

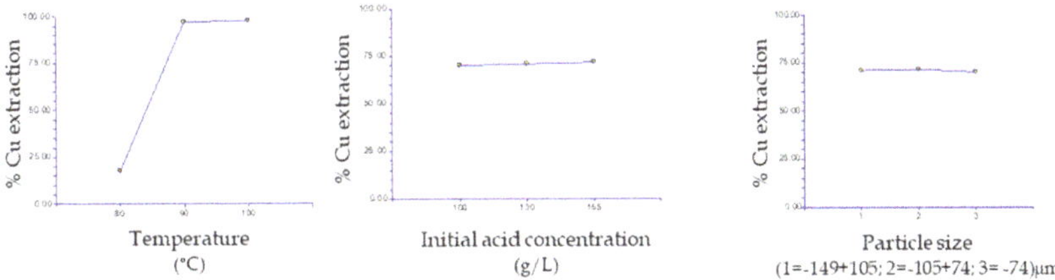

Figure 8. Correlations of temperature, initial acid concentration and particle size with extraction of copper.

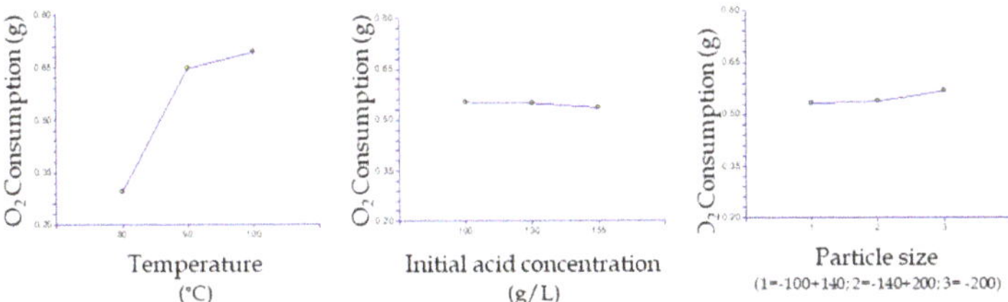

Figure 9. Correlations of temperature, initial acid concentration and particle size with oxygen consumption.

According to the above results the copper extraction can be calculated with the following multiple regression equation:

$$\text{Copper extraction } (\%) = -146.382 + 2.253\ T + 0.0957\ \text{Acid} - 0.0554\ \text{Size} \quad (4)$$

where T is the temperature expressed in °C; Acid is the initial acid concentration in g/L and size is the particle size of the chalcopyrite concentrate in µm.

3.4. *Effect of Temperature*

As observed in Figure 3, the different tests can be categorized into three groups with differing rates of reaction that were principally determined by temperature. The tests of the first group were carried out at 100 °C (tests 3, 5 and 7) and resulted in 97% copper extraction within 3 h. The second group (tests 2, 4 and 9) resulted in 55–77% copper extraction within 7 h. The tests of the final group were carried out at 80 °C (tests 1, 6 and 8) and resulted in 10–20% copper extraction within 7 h.

To highlight the required temperature for activating the decomposition of chalcopyrite concentrates, test 5 was replicated with a slowly-increasing temperature. Figure 10 shows the percentage of copper extraction and temperature versus time. As observed, the temperature had to reach 92–95 °C to decompose the chalcopyrite in the concentrate.

3.5. *Effect of Particle Size*

In the three groups that formed with respect to different reaction temperatures (Figure 3), the reactions were also faster for concentrates of small particle size; nevertheless, in the group that reacted at 100 °C, the −149 + 105 µm chalcopyrite concentrate reacted more rapidly than the concentrate filtered by −75 µm. The reactions in this group could have been slowed by the heat transfer from the jacket of the reactor to the suspension, resulting in different rates of reaction.

Thus, even when the chalcopyrite concentrate was a larger particle size (−149 + 105 µm), the copper extraction was not affected at 100 °C, and a similar level of extraction of copper was obtained.

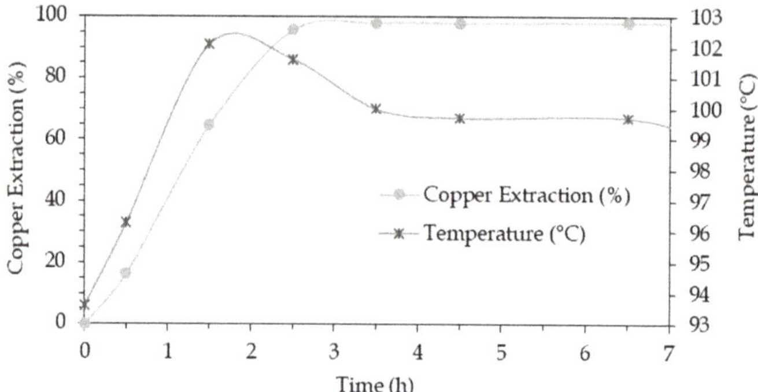

Figure 10. Copper extraction and temperature for the test 5 replicated.

3.6. *Effect of Acidity*

The initial sulfuric acid concentration in the suspension must be calculated based on the stoichiometry of the compounds that consume acid and the final acid concentration required to keep iron and copper in solution.

According to Figure 8, the initial sulfuric acid concentration is not significant for copper extraction. But in other exploratory tests carried out with the same copper concentrates at 100 °C under the same conditions, the suspension was found to require at least 15 g/L of sulfuric acid in solution to avoid the precipitation of iron as plumbojarosite in the residue. Thus, to prevent any problems in the recovery of valuable metals resulting from the presence of elemental sulfur and plumbojarosite in residues, and to avoid any potential impact on the profitability of operations, the initial sulfuric acid concentration is important to consider.

$$3Fe_2(SO_4)_3 + 12H_2O + PbSO_4 = 2Pb_{0.5}Fe_3(SO_4)_2(OH)_6 + 6H_2SO_4 \quad (5)$$

Figure 11 shows the iron and acid concentrations in solution, the percentage of plumbojarosite in the residue and the percentage of copper extraction versus time in an exploratory test. The initial acid concentration was 72 g/L, yet it diminished to 10–15 g/L. At 15 g/L of sulfuric acid in solution, the precipitation of plumbojarosite in the residue began, leading to a clear decrease in the iron in solution from 23.6 g/L to 15.8 g/L.

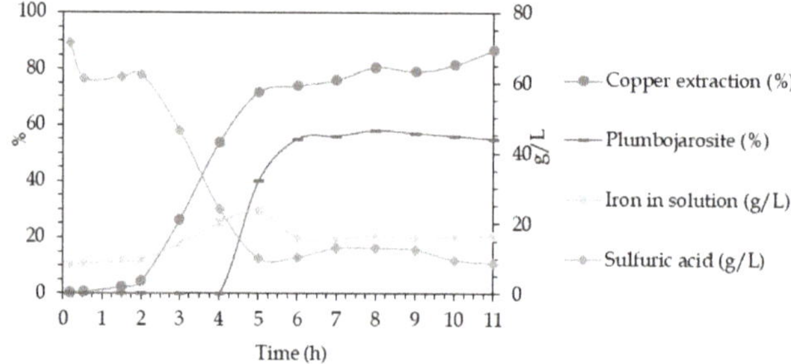

Figure 11. Results of an exploratory test with a low acid concentration in the reaction solution.

A comparison of the copper extractions with low and high acid concentrations in the reaction solution shows that passivation was promoted by a lack of acid in the solution, which, in turn, produced a plumbojarosite layer on the chalcopyrite surface (see Figure 12).

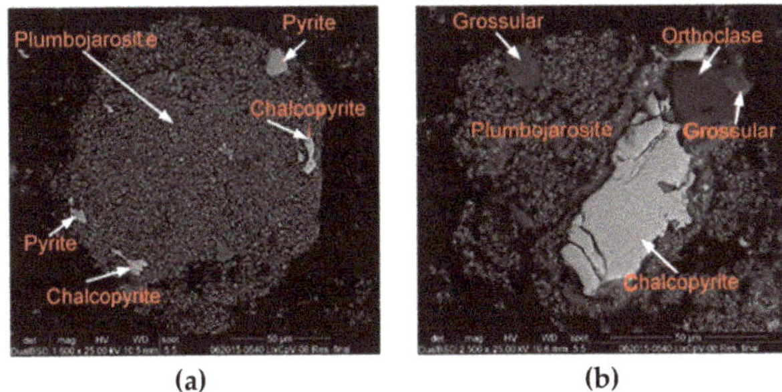

(a) (b)

Figure 12. Precipitation of plumbojarosite under low acidity conditions during leaching. (**a**) Plumbojarosite particle, (**b**) unreacted chalcopyrite particle.

The mass balance demonstrated that the ratio of sulfuric acid consumed (real) to that calculated by stoichiometry is given by Equation (6).

$$1.25 = \frac{\text{gReal}}{\text{g Stoichiometric} + \text{gtoreach } [\text{finalacid}] = 45\ \text{g/L}} \quad (6)$$

3.7. Kinetics

To determine the kinetics of the leaching process described herein, the shrinking core (product layer) model was applied to the real batch process considering the scanning electron microscopy (SEM) images of partly-reacted particles (Figure 4). The controlling step of the reaction was based on a comparison of the experimental data and assessment of which controlling model gives the best fit to the data. If the chemical reaction mechanism is assumed to be the controlling step, $1 - (1 - X)^{1/3}$ (X = conversion) is plotted as a function of time for the experimental data, and if the plot gives a linear correlation, the assumption is considered to be correct. Analogously, $1 - 3(1 - X)^{2/3}$ + $2(1 - X)$ (diffusion as controlling mechanism) can be plotted for the data when a non-porous product layer is formed [16,17].

The results show that the chemical reaction is the controlling stage for leaching copper from chalcopyrite concentrate. Figure 13 shows the linear regressions of the 9 tests. The equation of the chemical reaction as the controlling step is also shown as follows (Equation (7)).

$$kt = 1 - (1-X)^{(1/3)} \quad (7)$$

where t is time and k is the apparent velocity constant.

Table 10 shows the apparent velocity constants (k) and the determination coefficient (R^2) of the linear regression of the chemical reaction model for all 10 tests.

Table 10. Results for the linear regression of the chemical reaction model of all 10 tests.

Test	P1	P2	P3	P4	P5	P6	P7	P8	P9	P10
Slope (k)	0.006	0.032	0.12	0.031	0.096	0.009	0.112	0.01	0.035	0.0001
R^2	0.973	0.81	0.938	0.97	0.806	0.929	0.919	0.973	0.968	0.835

To calculate the activation energy, logarithms were applied to the Arrhenius equation (Equation (8)) to reformulate it as a linear equation. Accordingly, the logarithm of the apparent velocity constants versus the inverse of temperature of tests P3, P8, P9 and P10 is shown in Figure 14.

$$\text{Log}k = \text{LogA} - \log\,(E/R)^{\,(1/T)} \quad (8)$$

An activation energy of 61.93 kJ/mol was determined from the slope of the straight line in Figure 14. According to Habashi (1999), a chemically-controlled process is usually greater than 41.8 kJ/mol [18].

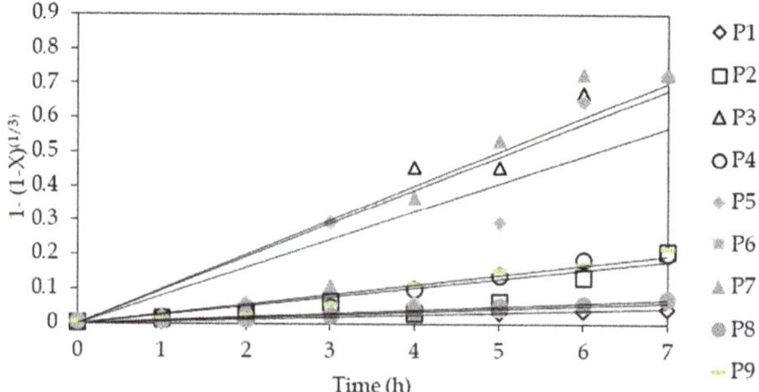

Figure 13. Linear regression of the chemical reaction model.

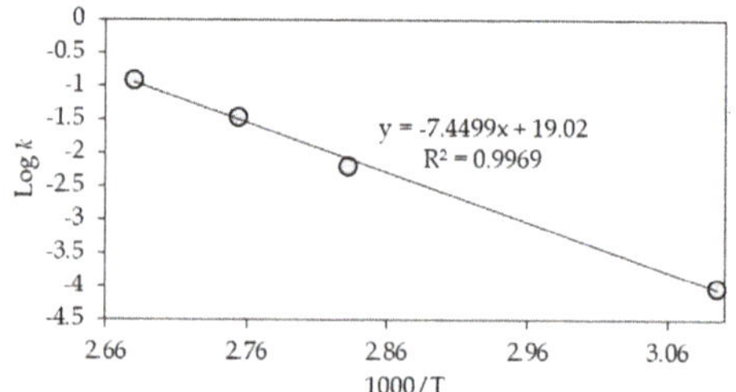

Figure 14. Logarithm of the apparent velocity constant versus the inverse of temperature.

In order to compare the activation energy with similar processes, Table 11 shows the results reported by some authors in literature. It can be observed that the activation energy for processes that use sulfuric acid, ferric ions and/or oxygen is similar to the obtained in this work that use sulfuric acid and oxygen. In the work reported by Padilla et al. (2008), they use also only sulfuric acid and oxygen in a pressure reactor and a copper extraction of 95% was obtained at an oxygen pressure of 5 kg/cm^2, 125 °C in 4 h [19]. In our work, we obtained 98% Cu extraction under less extreme conditions: oxygen pressure of 1 kg/cm^2, 100 C and only 3 h.

Table 11. Results reported in literature of activation energy for chalcopyrite leaching.

Leach Media	Activation Energy (kJ/mol)	Reference	Temperature Range (°C)
$K_2Cr_2O_7 + H_2SO_4$	24	[11]	50–97
$H_2O_2 + H_2SO_4$	39	[13]	30–80
$O_2 + H_2SO_4$	93.5	[19]	125–140
H_2SO_4	42.4	[20]	160–180
$K_2Cr_2O_7 + H_2SO_4$	48–54	[21]	-
$Fe_2(SO_4)_3 + H_2SO_4$	79.5	[22]	50–90
$H_2O_2 + H_2SO_4$	30	[23]	-
$Fe_2(SO_4)_3 + H_2SO_4$	21 ± 5	[24]	55–85
$NaNO_3 + H_2SO_4$	83	[25]	70–90
$Fe_2(SO_4)_3 + Cu^{2+} + NaCl + H_2SO_4$	66.6	[26]	70–90
$NaNO_2 + H_2SO_4$	34.0	[27]	80–120
$H_2SO_4 + O_2$	61.93	Present work	80–100

Padilla et al. (2008) required a higher activation energy than the present work; this is due to the design of reactor used. The reactor used in the present work promotes a high interaction between the solid-gas-liquid phases, improving the mass transport at the gas-solid interface. Thus, the activation energy required for the leaching of chalcopyrite decreases.

4. Conclusions

In the present study, the copper leaching of chalcopyrite concentrate in a 30-L batch reactor was described. The experimental results showed that it is possible to extract 98% of copper in only 3 h. This result indicates a fast process compared with others reported in literature.

The best result (98% in 3 h) was obtained under the following reaction conditions: 130 g/L of initial sulfuric acid concentration, temperature of 100 °C, oxygen pressure of 1 kg/cm^2, solid concentration of 100 g/L and concentrate particle size of −104 + 75 μm.

The copper leaching is controlled chemically. Then, the elemental sulfur layer exposed on the unreacted particles of chalcopyrite does not interfere with the mass transport or the interactions between phases.

A statistical analysis showed that temperature is the most important variable influencing the extraction of copper and oxygen consumption. A temperature of at least 92 °C (61.93 kJ/mol) is necessary to activate the decomposition of chalcopyrite.

The initial sulfuric acid concentration must also be considered as an important variable from an economic perspective. An excess of sulfuric acid will increase the neutralizing agent in the posterior stages of the leaching process, whereas a lack of sulfuric acid could result in the precipitation of iron as plumbojarosite, and could therefore create difficulties in the recovery of valuable metals at later stages.

Author Contributions: Conceptualization, J.C. and R.C.; Formal analysis, J.C.; Investigation, J.C., J.P., J.V. and R.C.; Methodology, J.C., J.V. and R.C.; Project administration, I.A.; Resources, J.P. and I.A.; Supervision, J.P., R.C. and I.A.; Validation, J.C. and I.A.; Writing—original draft, J.C.; Writing—review & editing, J.V. and R.C.

Acknowledgments: The authors gratefully acknowledge the technical support of the National Institute of Technology of México and Servicios Especializados Peñoles S.A. de C.V. for the financial support.

Conflicts of Interest: The authors declare no conflict of interest.

References

1. Taylor, A. *A-Z of Copper Ore Leaching Short Course Manual*; ALTA Metallurgical Services: Perth, Australia, 2014.
2. Eksteen, J.; Oraby, E.; Tanda, B. An alkaline glycine-based process for copper recovery and iron rejection from chalcopyrite. In Proceedings of the IMPC 2016—28th International Mineral Processing Congress, Québec City, QC, Canada, 11–15 September 2016; ISBN 9781926872292.
3. Hernández, P.; Taboada, M.; Herreros, O.; Graber, T.; Ghorbani, Y. Leaching of Chalcopyrite in Acidified Nitrate Using Seawater-Based Media. *Minerals* **2018**, *8*, 238. [CrossRef]
4. Watling, H. Chalcopyrite hydrometallurgy at atmospheric pressure: 2. Review of acidic chloride process options. *Hydrometallurgy* **2014**, *146*, 96–110. [CrossRef]
5. Turan, M.; Arslanoglu, H.; Altundogan, H. Optimization of the leaching conditions of chalcopyrite concentrate using ammonium persulfate in an autoclave system. *J. Taiwan Inst. Chem. Eng.* **2015**, *50*, 49–55. [CrossRef]
6. Frias, C.; Vera, E.; Romero, A.; Blanco, J. Flotation and hydroprocessing of bulk concentrate from Las Cruces Mine. In Proceedings of the Pb-Zn Conference, Düsseldorf, Germany, 14–17 June 2015.
7. Chen, S.; Berg, C.; Mansikkaviita, H.; Dezi, W.; Rong, Z. Drying Blended Concentrates by Kumera Steam Dryer for Kivcet Lead Flash Smelting. In Proceedings of the Pb-Zn Conference, Düsseldorf, Germany, 14–17 June 2015; pp. 257–265.
8. Burrows, A.; Azekenov, T.; Zatayev, R. Lead ISASMELTTM Operation at Ust-Kamenogorsk. In Proceedings of the Pb-Zn Conference, Düsseldorf, Germany, 14–17 June 2015; pp. 245–256.
9. Wang, C.; Yin, F.; Gao, W.; Ma, B. The research and Engineering Practice of the Lead Oxygen-Enrichment Flash Smelting. In Proceedings of the Pb-Zn Conference, Düsseldorf, Germany, 14–17 June 2015; pp. 273–284.
10. Outotec. *Software HSC Chemistry 8*; Continuous Research and Development: Pori, Finland, 2014.

11. Aydogan, S.; Ucar, G.; Canbazoglu, M. Dissolution kinetics of chalcopyrite in acidic potassium dichromate solution. *Hydrometallurgy* **2006**, *81*, 45–51. [CrossRef]
12. Xian, Y.; Wen, S.; Deng, J.; Liu, J.; Nie, Q. Leaching chalcopyrite with sodium chlorate in hydrochloric acid solution. *Can. Metall. Q.* **2012**, *51*, 133–140. [CrossRef]
13. Adebayo, A.; Ipinmoroti, K.; Ajayi, O. Dissolution kinetics of chalcopyrite with hydrogen peroxide in sulphuric acid medium. *Chem. Biochem. Eng. Q.* **2003**, *17*, 213–218.
14. Aleksei, K.; Kirill, K.; Stanislav, N. Pressure Leaching of Chalcopyrite Concentrate. *AIP Conf. Proc.* **2018**. [CrossRef]
15. O'brien, R.; Macdonald, C.; Meadows, N. *Chloride–Sulphate Leaching from Chalcopyrite Ores from Sabah. ALTA Copper Sulphides Symposium and Copper Hydrometallurgy Forum (GoldCoast, QLD)*; ALTA Metallurgical Services: Melbourne, Australia, 1999.
16. Levenspiel, O. *Chemical Reaction Engineering*, 3rd ed.; Wiley: New York, NY, USA, 1999; ISBN 9780471254249.
17. Grénman, H.; Salmi, T.; Murzin, D.Y. Solid-liquid reaction kinetics—Experimental aspects and model development. *Rev. Chem. Eng.* **2011**, *27*, 53–77. [CrossRef]
18. Habashi, F. *Kinetics of Metallurgical Process*; Métallurgie Extractive: Québec City, QC, Canada, 1999; ISBN 978-2-922686-60-9.
19. Padilla, R.; Pavez, P.; Ruiz, M. Kinetics of copper dissolution from sulfidized chalcopyrite at high pressures in H_2SO_4–O_2. *Hydrometallurgy* **2008**, *91*, 113–120. [CrossRef]
20. Baisui, H.; Batnasan, A.; Kazutoshi, H.; Yasushi, T.; Atsushi, S. Leaching and Kinetic Study on Pressure Oxidation of Chalcopyrite in H_2SO_4 Solution and the Effect of Pyrite on Chalcopyrite Leaching. *J. Sustain. Metall.* **2017**, *3*, 528–542. [CrossRef]
21. Antonijevic, M.; Jankovic, Z.; Dimitrijevic, M. Investigation of the kinetics of chalcopyrite oxidation by potassium dichromate. *Hydrometallurgy* **1994**, *35*, 187–201. [CrossRef]
22. Al-Harahsheh, M.; Kingman, S.; Hankins, N.; Somerfield, C.; Bradshaw, S.; Louw, W. The influence of microwaves on the leaching kinetics of chalcopyrite. *Miner. Eng.* **2005**, *18*, 1259–1268. [CrossRef]
23. Mahajan, V.; Misra, M.; Zhong, K.; Fuerstenau, M. Enhanced leaching of copper from chalcopyrite in hydrogen peroxide-glycol system. *Miner. Eng.* **2007**, *20*, 670–674. [CrossRef]
24. Kaplun, K.; Li, J.; Kawashima, N.; Gerson, A. Cu and Fe chalcopyrite leach activation energies and the effect of added Fe^{3+}. *Geochim. Cosmochim. Acta* **2011**, *75*, 5865–5878. [CrossRef]
25. Sokic, M.; Markovic, B.; Zivkovic, D. Kinetics of chalcopyrite leaching by sodium nitrate in sulphuric acid. *Hydrometallurgy* **2009**, *95*, 273–279. [CrossRef]
26. Veloso, T.; Peixoto, J.; Pereira, M.; Leao, V. Kinetics of chalcopyrite leaching in either ferric sulphate or cupric sulphate media in presence of NaCl. *Int. J. Miner. Process.* **2016**, *148*, 147–154. [CrossRef]
27. Gok, O.; Anderson, C.; Cicekli, G.; Cocen, E. Leaching kinetics of copper from chalcopyrite concentrate in nitrous-sulfuric acid. *Physicochem. Probl. Miner. Process.* **2013**, *50*, 339–413. [CrossRef]

Article

A Kinetic Study on the Preparation of AlNi Alloys by Aluminothermic Reduction of NiO Powders

Cesar Silva Beltran [1], Alfredo Flores Valdes [1], Jesús Torres Torres [1] and Rocio Ochoa Palacios [2,*]

1 Centro de Investigación y de Estudios Avanzados del IPN, Unidad Saltillo Avenida Industria Metalúrgica 1062, Parque Industrial Saltillo-Ramos Arizpe, 25900 Ramos Arizpe, Coahuila, Mexico; slash_sil@hotmail.com (C.S.B.); alfredo.flores@cinvestav.edu.mx (A.F.V.); jesus.torres@cinvestav.edu.mx (J.T.T.)

2 Instituto Tecnológico de México, Campus Saltillo, Venustiano Carranza, 25000 Saltillo, Coahuila, Mexico

* Correspondence: rochoa@itsaltillo.edu.mx; Tel.: +52-844-438-9500

Received: 9 July 2018; Accepted: 31 July 2018; Published: 28 August 2018

Abstract: In this work, the experimental results obtained during the preparation of Al-Ni and Al-Ni-Mg alloys using the aluminothermic reduction of NiO by submerged powder injection, assisted with mechanical agitation are presented and discussed. The analyzed variables were melt temperature, agitation speed, and initial magnesium concentration in the molten alloy. For some of the experiments performed, it was found that the Ni concentration increased from 0 to about 3 wt-% after 90 min of treatment at constant temperature and constant agitation speed. In order to determine the values of the kinetic parameters of interest, such as the activation energy and the rate constants, the values of the results obtained were fitted to the kinetic formulae available. Moreover, the kinetics of the reaction were found to be governed by the diffusion of Al and Mg to the NiO boundary layer, where $MgAl_2O_4$ or Al_2O_3 were formed as the main reaction products. Finally, from a thermodynamic study of the system, the main reactions that took place are explained.

Keywords: Al-Ni alloys; aluminothermic reactions; reaction rate; Al master alloys; kinetics

---

1. Introduction

Aluminum-nickel alloys are widely used in the manufacturing of parts for the automotive industry, particularly those with a high nickel content, around 3 wt-%, and in cutting tools with TiAl because of their high resistance and low corrosion [1–4]. It is expected that in the upcoming years, the production of this type of alloy will increase worldwide, to reach a level of production higher than 400,000 tons per year. The possibility of preparing Al-Ni and Al-Ni-Mg alloys from the aluminothermic reduction of NiO has been proposed, using the submerged injection of powders as a technique for incorporating the nickel oxide particles.

In this sense, the first applications of the aluminothermic reduction of oxides were in the preparation of molten iron streams for filling the enclosures of exposed surfaces, where the slag rich in Al_2O_3 could be separated easily because of its lower density relative to that of iron [5]. Currently, an aluminothermic reduction is used for the preparation of aluminum master alloys, such as Al-Sr-Mg, Al-Ce-Mg, Al-Zn, Al-Mn, Al-Zr, Al-Mg-Fe-Cr, Al-Li, and so on. [6–15]. Nowadays, aluminothermic reduction is the most common practice used in the production of both kinds of products, pure metals or alloys [16–19]. However, only recently has the understanding of the kinetics of reaction been addressed for the systems of reaction, as every system is quite different.

The reduction reaction of nickel oxide by molten aluminum has a considerably negative value for the standard Gibbs free energy of reaction at operating temperatures, as is shown by the next reaction:

$$3MO + 2Al = Al_2O_3 + 3M \quad (1)$$

In the specific case of NiO, it is possible to accelerate the reduction reaction of nickel oxide with magnesium, thereby increasing its concentration in the molten aluminum during the aluminothermic process. In this case, the aluminothermic reaction occurring at 1073 K could be as follows:

$$2Al + Mg + 4NiO = MgAl_2O_4 + 4Ni,\ \Delta G^{\circ}{}_{1073\ K} = -869.48\ kJ \quad (2)$$

Magnesium enhances the reaction rate, not only through its effect on the chemical reactivity of the molten bath, but also through its effect on the surface tension of molten aluminum, improving the wettability between the NiO particles and the molten aluminum solution [20].

On the other hand, it is broadly known that in the last 30 years, a lot of polluting waste from secondary spent batteries has been generated, of which NiO and nickel hydroxide (NiOH) are the main components. As such, it is attractive to investigate the aluminothermic reduction rate of NiO to formulate and propose alternatives for the use of materials that come from the recycling industry, and that can have a high added value, such as the preparation of Al-Ni or Al-Ni-Mg type alloys. In this sense, the main purpose of this paper is to present and discuss the results of an investigation aimed at understanding the role of temperature, initial magnesium concentration in molten aluminum, and the degree of agitation during the preparation of Al-Ni and Al-Ni-Mg alloys using a submerged powder injection of NiO powders, assisted by mechanical agitation. The results mainly focus on measuring the reaction rate and values of some of the kinetic parameters of interest, such as the rate constants and activation energy values of the process, where describing the mechanism of the reaction is of paramount importance.

2. Materials and Methods

A 70 KW Power Track 75–30 electromagnetic medium frequency induction furnace, equipped with a silicon carbide crucible with a capacity of 10 kg of molten aluminum, was used as the melting unit. The powder injection equipment, whose scheme is presented in Figure 1, allowed for the continuous and controlled feeding of the solid material (NiO) through an inert carrier gas, which in this investigation, was high purity argon (99.99%). The design of the powder injection equipment used for the tests basically consisted of a pressurized chamber, inside which was a cylindrical container to deposit the reactive powder. This container was provided with a worm on the inside, which was rotated by an electric motor connected to a voltage controller located on the control board. This allowed us to vary the speed of rotation of the worm, and therefore, control the speed of the feeding of the powder.

The mixture of powder/gas falls directly into a funnel that connected the feeding line to the outlet lance of the pressurized system, being dragged thereto by means of the carrier gas. In this way, a powder/gas mixture is conducted towards the melting unit, where it is injected inside the molten metal through a graphite lance. This lance is located at a depth of 85% of the height of the molten metal. The location of the injection lance with respect to the geometry of the furnace, plays a very important role in the reaction kinetics, facilitating the discharge of the powder, therefore assuring its contact with the molten metal solution. The dimensions of this lance were as follows: length, 50 cm; external diameter, 5.08 cm; and diameter of the internal hole, 3.54 cm. To promote a better agitation inside the furnace, a graphite agitator was built, which was assembled to a mechanical drill and placed in the center of the bath; in this way, the agitation was constant and vigorous, as, in addition to the agitator and the gas injected, the Eddy electromagnetic waves generated by the induction coil helped to reach a greater efficiency in the mixing conditions. An insulating lid was attached to the furnace to minimize the inlet of air to the furnace. This stage had an entry for the mechanical agitator and another for the injection lance.

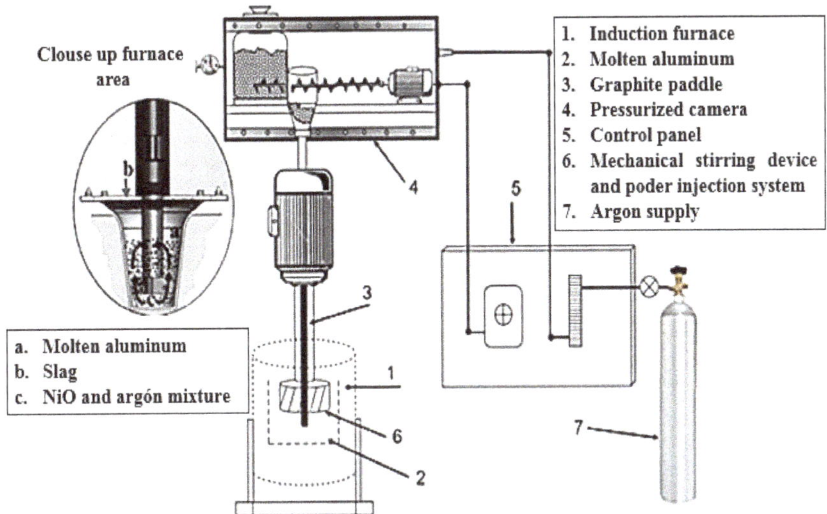

Figure 1. Schematic illustration of the experimental set-up employed in the experiments of the submerged powder injection with mechanical agitation.

Once the lid and the agitator were placed, the temperature and the revolutions per minute were adjusted while the lance was inserted in the molten aluminum to later adjust the powder feeding speed, thus initiating the treatment. We conducted 27 experiments, where the effects of temperature, percentage of magnesium in the alloy, and agitation speed in the molten bath were investigated. The operational parameters and their values were selected from preliminary works [8–12], where the kinetics of the aluminothermic reactions have been studied, not only for attaining reliable reaction efficiencies, but also to avoid metal losses during treatment. Initially, a factorial design was proposed to reduce the experimental errors. The factorial design was comprised of three factors and three levels, as is shown in Table 1. The number of experiments was performed according to the data in Table 1 and Equation (3).

Table 1. Parameters and levels of the initial experimental design.

Factors	Level		
	1	2	3
Temperature (K)	1023	1073	1123
Mg (wt-%)	0	2	3
Agitation (rpm)	50	100	300

$$N = a \times b \times c \times n \tag{3}$$

where N = numbers of experiments; a = 3 (factor A levels) A level; b = 3 (factor B levels) B level; c = 3 (factor C levels) C level; and n = 1 (number of replicas).

According to this, the total number of experiments (E_i) is shown in Table 2. To calculate the number of reagents to be injected, the amount of aluminum, the nickel amount released from NiO, and the stoichiometry given by Reaction (2) were considered. Therefore, the amount of theoretical NiO that could feasibly be added to the molten bath was calculated as 250 g, also by considering the magnesium concentration in the corresponding trial tests that required it. The target Ni concentration was 3 wt-%. The parameters that remained constant were the initial quantity of molten aluminum,

5 kg; powder injection rate, 250 g min^{-1} of NiO; argon injection rate, 5 L min^{-1}; and particle size of the NiO powders, to the order of a 4 μm average.

Table 2. Design of total experiments ($E_{i,j,k}$). *i*—temperature (K); *j*—initial Mg concentration (wt-%); *k*—agitation speed (rpm).

E1 1023,0,50	E2 1023,0,100	E3 1023,0,300	E4 1073,2,50	E5 1073,2,100
E6 1073,2,300	E7 1073,3,50	E8 1073,3,100	E9 1073,3,300	E10 1073,0,50
E11 1073,0,100	E12 1073,0,300	E13 1073,2,50	E14 1073,2,100	E15 1073,2,300
E16 1073,3,50	E17 1073,3,100	E18 1073,3,300	E19 1123,0,50	E20 1123,0,100
E21 1123,0,300	E22 1123,2,50	E23 1123,2,100	E24 1123,2,300	E25 1123,3,50
E26 1123,3,100	E27 1123,3,30			

Samples from the molten bath were taken every 10 min, attaining up to 90 min of treatment. These samples were poured into a metallic mold and marked accordingly for chemical analysis determinations and microscopic observations, using both optical and scanning electron microscopy (SEM). The response variables were the actual nickel and magnesium concentrations, and the amount of Ni or Ni-Mg rich phases as a function of the injection time. A chemical analysis was carried out using the spark emission spectrometry technique. The chemical composition of the aluminum alloy is given in Table 3, while the chemical composition of the NiO powders is given Table 4.

Table 3. Chemical composition of the aluminum alloy (wt-%).

Element	Si	Fe	Cu	Mn	Mg	Cr	Ni	Zn	Sn	Ti	Al
wt-%	0.013	0.48	0.025	0.012	0.089	0.0012	0.0003	0.021	0.013	0.0012	99.34

Table 4. Chemical composition of the NiO powders (wt-%).

Element	Ni	Co	Fe	Cu	Zn	Mn	Mg	Ca	Na	S	O
wt-%	76.39	0.02	0.005	0.001	0.007	0.02	0.01	0.009	0.004	0.06	23.45

At the end of each experiment, the samples from the produced slags were taken for characterization by X-ray diffraction. To determine the Standard Gibbs free energy of reaction, the software HSC [21] was used, while for the drawing diagrams of the phase stability as a function of the temperature of the reaction, the software Factsage Chemistry Software was used [22].

3. Results and Discussion

3.1. Experimental Results

Figures 2 and 3 show that the Ni concentration increased in the molten alloys as a function of the injection time for the temperatures and initial magnesium concentrations indicated, at the agitation speed of 300 rpm. From these graphs, the nickel concentration reached the highest value at higher magnesium concentrations, at the temperature of 1123 K.

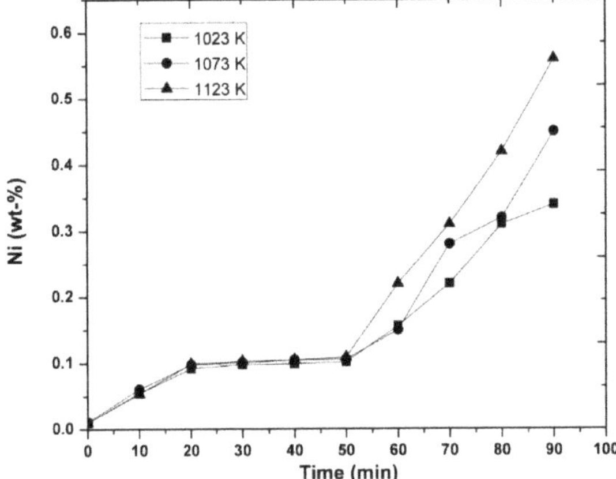

Figure 2. Increase in nickel concentration as a function of injection time at the indicated temperatures, for 0 wt-% Mg, and a constant agitation speed of 300 rpm.

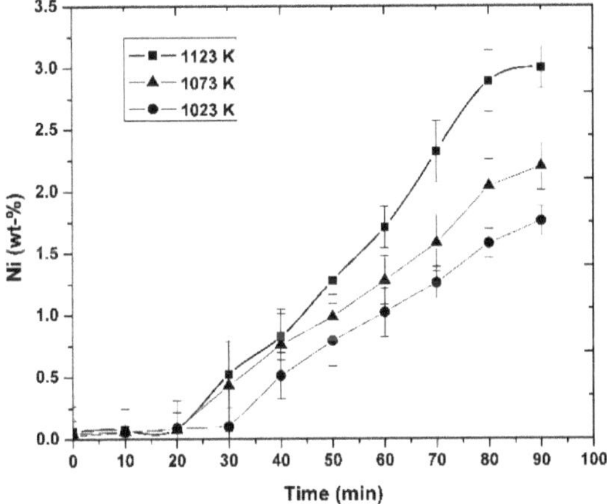

Figure 3. Increase in the nickel concentration as a function of injection time at the indicated temperatures for 3 wt-% Mg content in the alloy and at a constant agitation speed of 300 rpm.

Figure 4 shows the increase in the Ni concentration as a function of the injection time for the agitation speeds considered, at the constant initial Mg concentration and temperature indicated. From this figure, it can be observed how the agitation speed plays an important role in the aluminothermic reduction of the NiO particles, which was due to the improved mass transfer conditions attained at higher values of this parameter. Indirectly, it also indicated the diffusive nature of the process taking place.

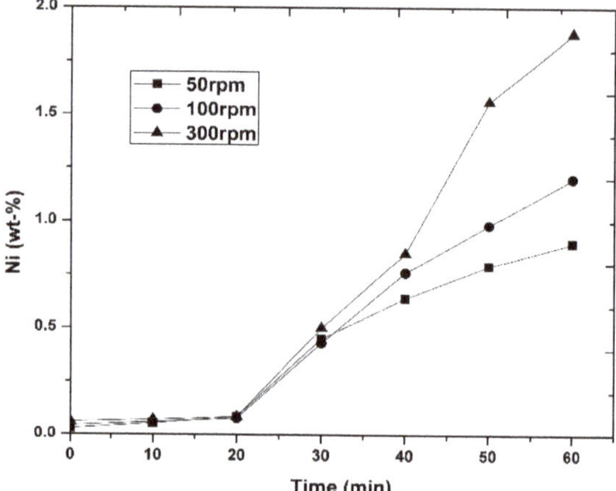

Figure 4. Increase in the nickel concentration as a function of injection time, at the indicated agitation speeds, for an initial Mg concentration of 3 wt-%, and a temperature of 1123 K.

From the explanations of Guedes et al. [23], the increase in the nickel concentration as a function of the injection time, as shown in Figures 2 and 3, can be explained as follows. When magnesium is not present in the molten alloy, a layer of Al_2O_3 is formed at the beginning of the NiO reduction reaction. The growth of this layer minimizes the contact between the molten aluminum and the reaction front, acting as a barrier between both the aluminum and nickel oxide, which prevents the reaction given by Equation (2) to be completed. However, the Al_2O_3 layer is not completely impervious, thus having some porosity through which the aluminum can diffuse to the boundary layer, where it reaches the surface of the NiO particles. Magnesium improves the reaction rate in two ways. At 973 K, it reduces the surface tension of the molten aluminum from 0.91 (with no Mg present) to 0.66 N m^{-1} for a magnesium concentration equal to 1 wt-% [24]. This facilitates the wettability of the NiO particles in the molten phase. On the other hand, Mg increases the reactivity of the bath, because this element also reduces the NiO to 973 K, according to Reaction (2), where the greater negative value of its Gibbs free energy indicates the spontaneity of this reaction.

During their experiments on the synthesis of the Al/Al_2O_3 composites, Pai and Ray [25] found that MgO could be formed from the reduction of Al_2O_3 by the magnesium contained in the alloy. This reduction reaction can also occur in the case of the NiO reduction reaction given by Equation (2). In turn, MgO and Al_2O_3 can react during solidification to form $MgAl_2O_4$ as the final reaction product, according to the following reaction between pure compounds, whose ΔG° is calculated at 1073 K.

$$MgO + Al_2O_3 \rightarrow MgAl_2O_4,\ \Delta G^{\circ}_{1073\ K} = -92.02\ kJ \quad (4)$$

The successive formation of MgO and Al_2O_3 causes expansion and contraction during the reaction, which results in breaking the crystals into numerous smaller crystals, thus producing the porosity required for the diffusion of aluminum, magnesium onwards, or nickel backwards.

Figure 5 shows the variation in the magnesium concentration as a function of the NiO injection time, and the concentration of magnesium decreased continuously during the injection, reaching the lower value of 1.98 wt-% at the end of the experiments for the given conditions established.

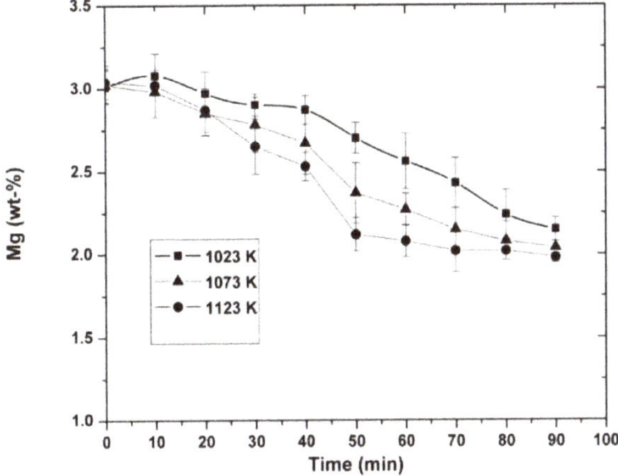

Figure 5. Decrease of the magnesium concentration in molten aluminum as a function of injection time at the indicated temperatures at the constant agitation speed of 300 rpm.

The micrographs in Figure 6 show the evolution of the microstructure of the aluminothermic reduction reaction of the NiO powders as a function of the indicated injection times from samples obtained at 1123 K, at a constant agitation speed of 300 rpm. In these micrographs, it was observed that the amount of the nickel-rich intermetallic phase increased as the treatment time increased, therefore the amount of nickel incorporated in the molten alloy increased with increasing time.

Figure 7 shows the increase in the amount of Ni-rich intermetallic phase as a function of the injection time, hence, the amount of nickel incorporated was increased in the molten alloy by the aluminothermic reduction reaction of the NiO powders.

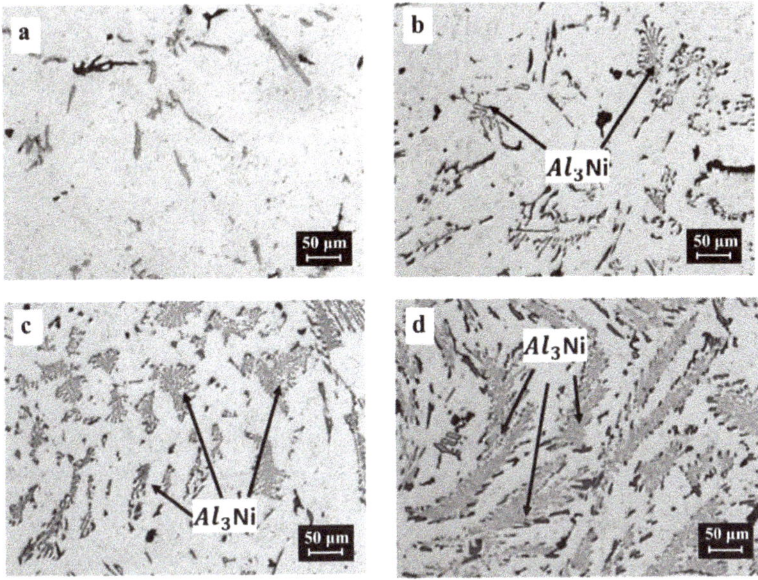

Figure 6. Photomicrographs of Al-Ni samples obtained at 1123 K, as a function of injection time, where (**a**) 0 min, (**b**) 20 min, (**c**) 40 min, and (**d**) 90 min.

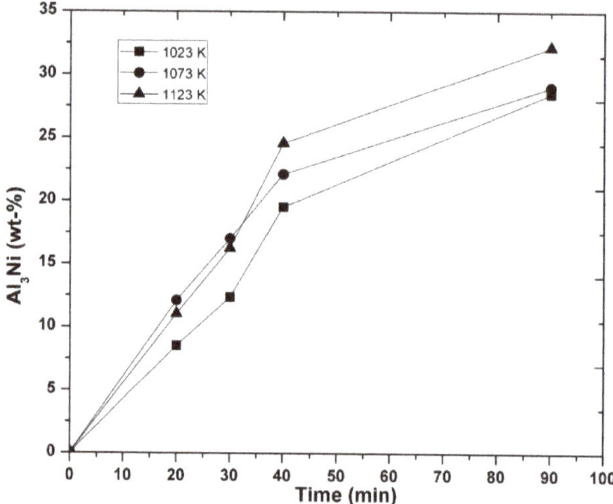

Figure 7. Increase of the Al_3Ni intermetallic phase as a function of injection time for the temperatures indicated at the constant agitation speed of 300 rpm.

The micrographs in Figure 8 show the evolution of the thickness of the layers of the NiO particles reacted as a function of the reaction time, taken from the molten metal at a temperature of 1123 K for an agitation speed of 300 rpm. After 30 min, the particle remained unreacted (Figure 8a), and after 40 min of addition, it began to form an Al_2O_3 layer around the NiO particle (Figure 8b). This layer grew progressively until the reaction stopped because of a change in the mechanism that controls the reaction (Figure 8c,d). After 70 min of contact (Figure 8e), the Al_2O_3 layer began to disappear. In the micrographs, it can be seen that after 90 min of reaction, the Al_2O_3 layer completely disappeared, leaving only the solid particle and the Al_3Ni intermetallic particles around them.

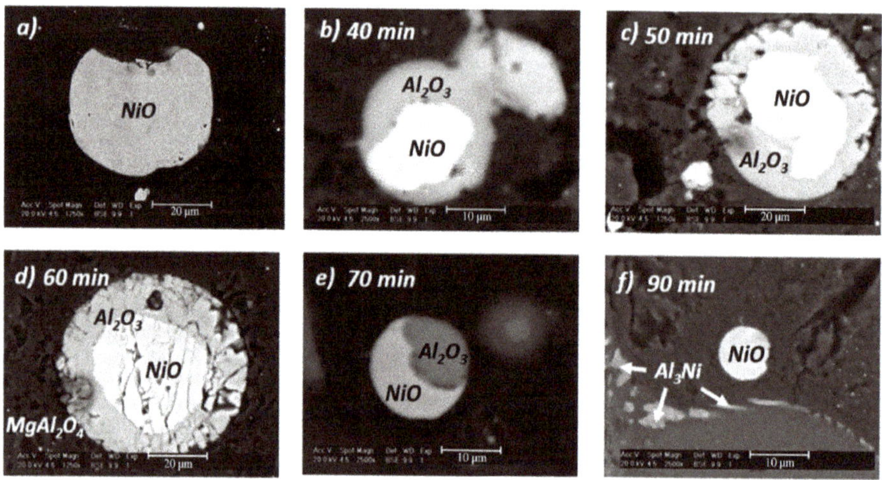

Figure 8. SEM images of partially reacted NiO particles obtained from molten aluminum at 1123 K at a stirring speed of 300 rpm for the indicated times of (**a**) 30 min, (**b**) 40 min, (**c**) 50 min, (**d**) 60 min, (**e**) 70 min, and (**f**) 90 min.

The particle shown in Figure 8d was analyzed separately by energy dispersion spectroscopy in the SEM, and the corresponding energy dispersive X-ray spectroscopy (EDS) patterns, together with the micrographs showing the microareas from where they were obtained, are shown in Figure 9. The microanalysis measurements show that the nucleus of NiO was surrounded by an Mg and Al-rich phase, marked as (2) and (3). The semi-quantitative chemical composition of the layer of the reaction product, as shown in Table 5, was observed as the phase formed was spinel ($MgAl_2O_4$). These EDS results suggest that the NiO was reduced by aluminum and magnesium, forming intermediate reaction products such as the spinel phase, while the Ni-rich phase corresponded to Al_3Ni [26].

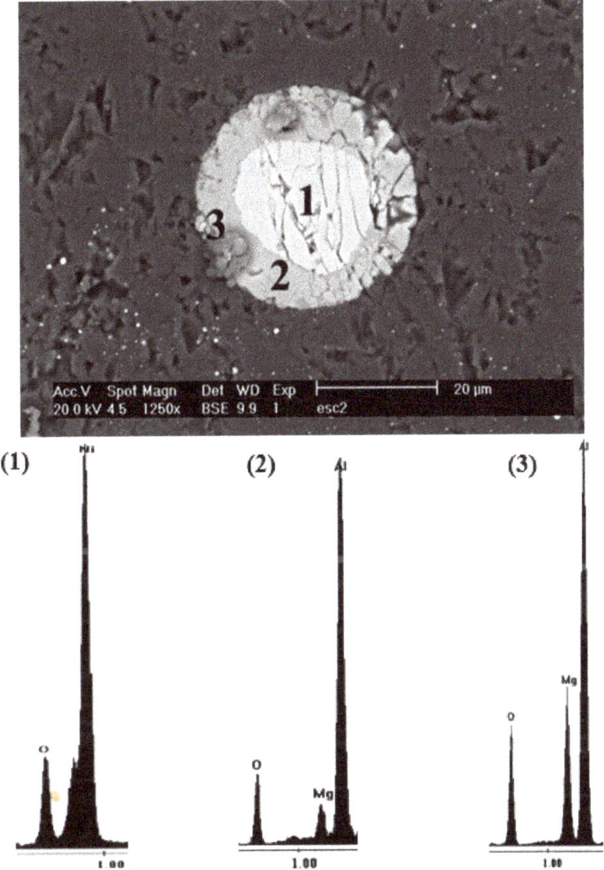

Figure 9. SEM images of the partially reacted NiO particle shown in Figure 8d, and the corresponding EDS patterns of the microareas of the (**1**) center, (**2**) outside layer, and (**3**) $MgAl_2O_4$.

Table 5. EDS chemical analysis of the partially reacted NiO particle shown in Figure 8d.

Particle	wt-%			
	Ni	O	Al	Mg
1	67.51	32.49	0	0
2	0	25.76	72.14	2.1
3	0	31.45	49.79	18.76

Figure 10 shows the X-ray powder diffraction (XRD) pattern corresponding to the slag obtained after the end of experiment E27, where there was the presence of MgO, $MgAl_2O_4$, aluminum, and NiO. The aluminum came from the molten alloy, as the sample was taken from a pasty area formed between the slag and molten aluminum.

The MgO and $MgAl_2O_4$ in the slag suggests that both aluminum and magnesium reduced the nickel oxide by a metallothermic mechanism. The NiO came from the partially reacted particles. The analysis of the slag helped to corroborate that the main aluminothermic reduction reaction was that given by Equation (2), where the Ni was obtained from NiO dissolved in the molten alloy to solidify as Al_3Ni particles, which are shown in Figure 5.

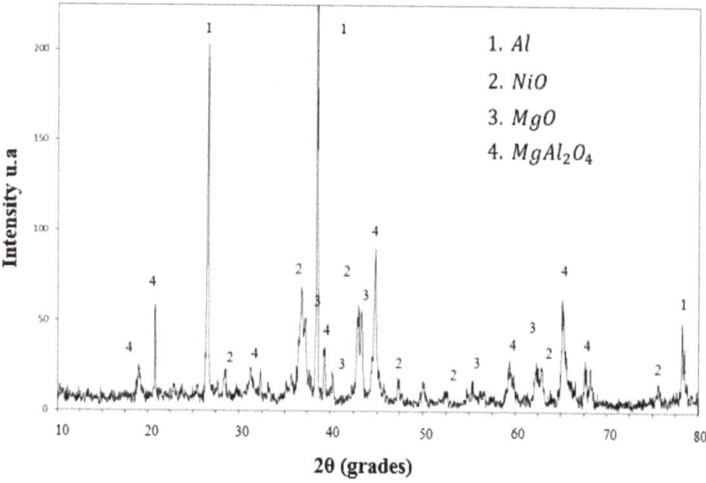

Figure 10. XRD pattern of the slag samples for an alloy where the initial magnesium was 3 wt-% at a stirring speed of 300 rpm and a temperature of 1123 K.

3.2. *Thermodynamic Consideration for the Al-NiO-Mg System*

The aluminum alloys obtained in this work contained between 2 to 3 wt-% of both Mg and Ni, therefore, a thermodynamic explanation would help to understand the expected results. During the aluminothermic reduction process, the Al, Mg, and NiO species are involved, and react to obtain different reaction products. The software FactSage allowed us to consider ideal solutions for the Al-Mg-Ni system in the solid and liquid states. The reactants that were introduced into the software were Al, Mg, and Ni as pure species. The balance of the system was computerized for the temperature range from 373 to 1273 K, where the activities considered for Al, Ni, and Mg were equal to 1. Figure 11 shows the results of the software in the Al-Mg-Ni equilibrium system. The phases that were present below the melting temperature of aluminum were Al(s), MgO, and Al_3Ni in their stable forms, while at temperatures higher than 933 K, the Al and Mg were in a liquid state. These results define the reactions involved during the reduction of NiO powders by Al-Mg(l). The formation of MgO is thermodynamically feasible from room temperature up to the melting temperature of aluminum. However, the diagram drawn by the software was preliminary as the kinetic conditions change continuously during the reaction.

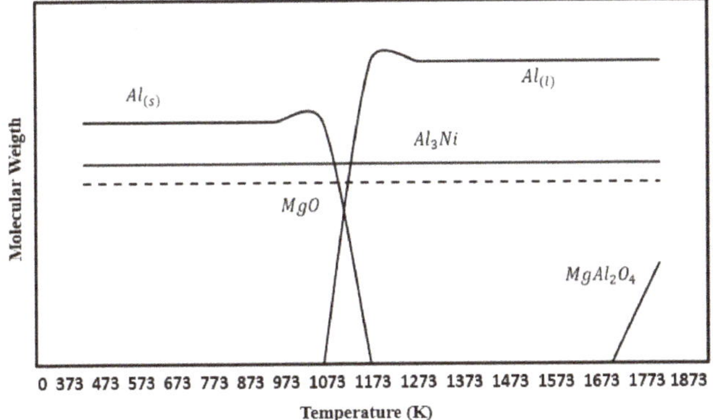

Figure 11. Equilibrium diagram of species for the Al-Mg-NiO system in the temperature range from 373 to 1873 K.

The reactions are presented and the values of ΔG° at 1123 K were obtained from the software, expressed per mole of oxygen.

$$4/3Al + O_2 = 2/3Al_2O_3,\ G^\circ_{1123\ K} = -882.46\ kJ \tag{5}$$

Reaction (5) can occur between aluminum and oxygen from the environment with pressures of 1 atm and elevated temperatures, forming Al_2O_3. The next reaction can occur between Al_2O_3 with magnesium and oxygen to form the spinel $MgAl_2O_4$, which is possible with amounts higher than 0.05% Mg.

$$Al_2O_3 + Mg + 1/2O_2 = MgAl_2O_4,\ \Delta G^\circ_{1123\ K} = -1036.56\ kJ \tag{6}$$

Reactions (5) and (6) must be produced at the aluminum/slag interface, forming a protective layer of aluminum and magnesium oxides in the molten bath.

The reaction between Al and NiO to form Al_2O_3 while releasing Ni is quite possible because of the action masses law, as aluminum is the element present in greater quantity in the system. The reaction is that given by Equation (1). However, the graphs in Figure 2 show that in the experiments performed with pure aluminum, the concentration of nickel attained was rather low. It was shown that Al_2O_3 [27] in a wide range of temperatures was not wettable by aluminum, because it has a contact angle value >90.5° [28].

Therefore, it is necessary to decrease the contact angle of Al_2O_3 by molten aluminum. According to the literature [29], one way to change the contact angle is to lower the surface energy of molten aluminum by adding Mg, as this is one of the elements that has this effect.

In the experiments that were carried out, it was observed that the magnesium content decreased with the increased time, as is shown in Figure 5. This was attributed not only to the effect of this element on the contact angle, but also to the high reactivity that magnesium possesses, according to the following reaction:

$$NiO + Mg = MgO + Ni,\ \Delta G^\circ_{1123\ K} = -342.83\ kJ \tag{7}$$

The global chemical reaction taking place is that given by Equation (2). The following reaction is possible because of the greater chemical reactivity of Mg than Al:

$$Al_2O_3 + 3Mg = 3MgO + 2Al,\ \Delta G^\circ_{1123\ K} = -117.69\ kJ \tag{8}$$

The formation of MgO, a phase present in the slag produced in this work for aluminum alloys with initial Mg concentrations above 2 wt-%, proceeded from the following reaction:

$$2Mg + O_2 = 2MgO,\ \Delta G^{\circ}_{1123\ K} = -970.09\ kJ \tag{9}$$

In turn, Al_2O_3 and MgO react according to Equation (4) to form the so-called spinel, $MgAl_2O_4$.

All of the reaction products described by the reactions given by Equations (1), (2), (4), (5), or (7)–(9) were found in the slags obtained from the NiO reduction process studied in this work.

3.3. Kinetic Measurements

The experimental evidence strongly indicated that the reaction rate in the studied process was controlled by the diffusion of chemical species, mainly Al or Mg, to the boundary layer, where they reacted with NiO. Consequently, the experimental information obtained at each temperature, for alloys with and without magnesium, was analyzed using the kinetic formulas presented by Brown [30]. However, to adapt to the experimental information of the available kinetic models, it was assumed that the observed nickel concentration corresponded to that of the complete reaction, and therefore, the final reacted fraction was equal to 1. First of all, it is necessary to state that for the present case, the reaction rate is defined as the ratio of the reacted fraction (α) against time (t), given by the following:

$$reaction\ rate = \frac{d\alpha}{dt} \tag{10}$$

As an associated magnitude to the progress of the aluminothermic reaction, α is numerically equal to the reacted fraction attained in a given period, called actual time. Therefore, reacted fraction, in our case α_{Ni}, is given by next equation:

$$\alpha_{Ni} = \frac{w_o - w_i}{w_o - w_f} \tag{11}$$

where w_o, w_i, and w_f are the *Ni* concentrations measured in the alloy at the beginning of the injection time, the actual concentration, and the final concentration, respectively.

Figures 12–14 show the change in the reacted fraction as a function of time, for the different set of parameters studied, resulted from the application of Equation (11) to the experimental results obtained. In general, it can be observed from these graphs that the aluminothermic reduction reaction was not completed for any value of the parameters studied (i.e., temperature, initial magnesium concentration, or agitation speed), owing to the diffusive nature of the reaction, as it will be explained later.

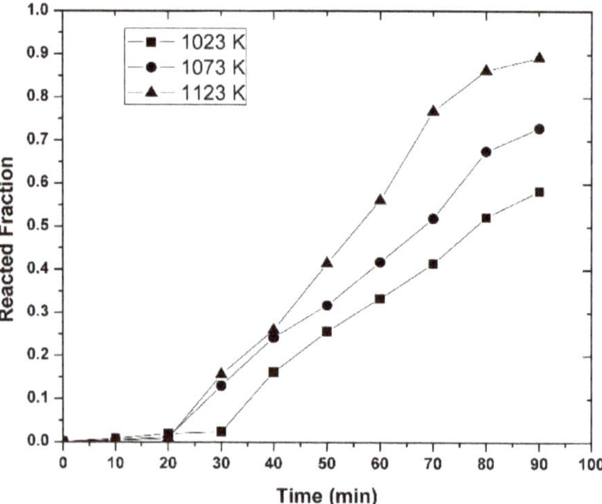

Figure 12. Reacted fraction of Ni as a function of time for the temperatures indicated, for a constant initial concentration of Mg of 3 wt-%, at the constant agitation speed of 300 rpm.

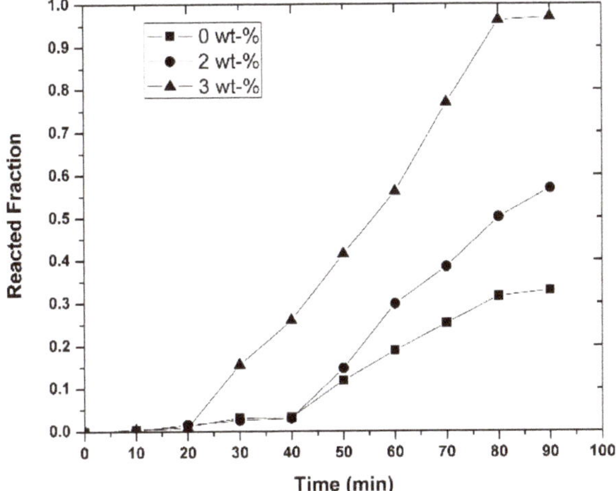

Figure 13. Reacted fraction of Ni as a function of time for the initial magnesium concentrations indicated, at the constant temperature of 1123 K, and constant agitation speed of 300 rpm.

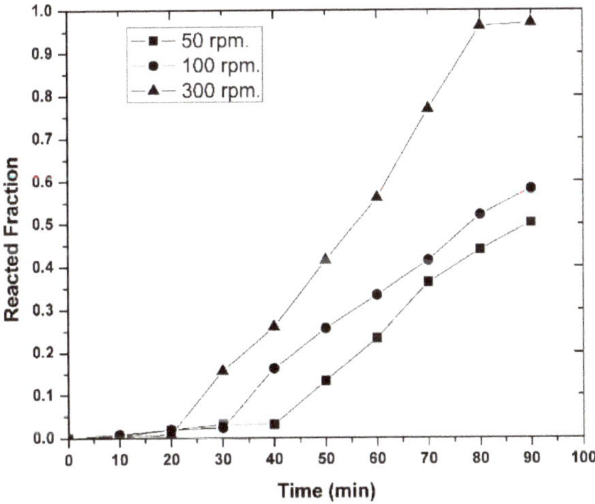

Figure 14. Reacted fraction of Ni as a function of time for the agitation speed indicated, for a constant initial concentration of Mg of 3 wt-%, at the constant temperature of 1123 K.

In this sense, it was found that the kinetic information was well suited for the diffusion model equation (D1), as shown in Figure 15, which corresponded to the experiments where alloys with magnesium were used at an agitation speed of 300 rpm. In turn, Figure 16 shows the experimental results at different temperatures. The equation of the model is as follows:

$$(1-\alpha)ln(1-\alpha)+\alpha \quad (12)$$

where α is the reacted fraction.

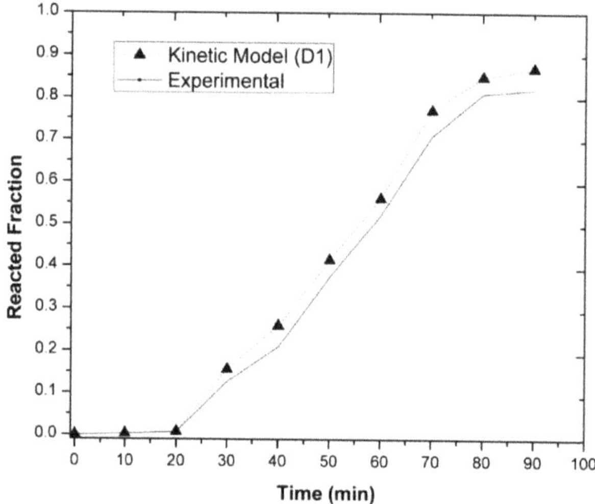

Figure 15. Experimental results and prediction of model D1 at the temperature of 1123 K for an alloy with 3 wt-% Mg at 300 rpm.

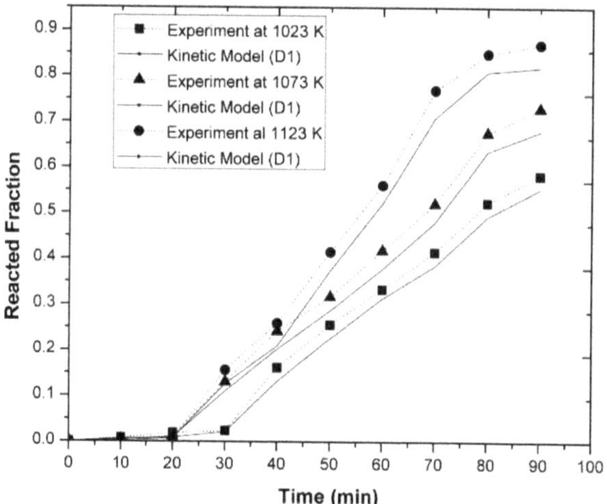

Figure 16. Experimental results and prediction of the D1 model at different temperatures for alloys with an initial Mg concentration of 3 wt-%, at 300 rpm.

The linear regression of *ln* (*k*) against $1/T$ was plotted to obtain the values of the activation energy and the rate constants using the Arrhenius equation given below:

$$k = k_0 e^{-E/RT} \tag{13}$$

where E is the activation energy (J mol^{-1}), and R is the universal gas constant (8.314 J mol^{-1} K). Figure 17 shows the linear dependence of a graph of *ln* (*k*) against $1/T$ for the aluminothermic reduction of NiO for alloys with magnesium (a) and without magnesium (b). From the slope values of these graphs, the values of the activation energies were determined as E = 35.75 KJ mol^{-1} for the alloys without magnesium and E = 15.80 KJ mol^{-1} for the alloys with an initial magnesium concentration of 3 wt-% Mg.

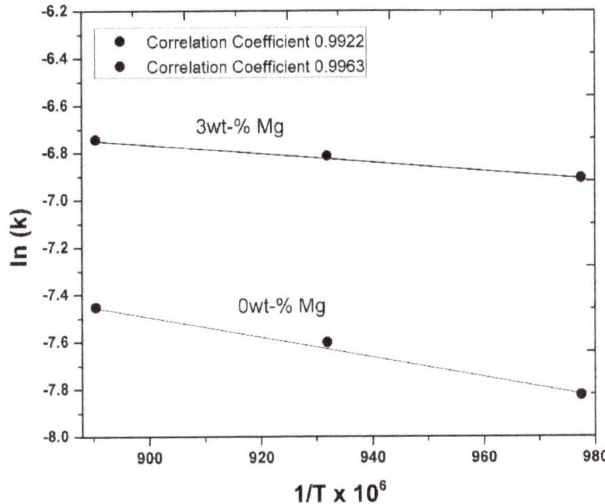

Figure 17. Graphs of *ln* (*k*) against 1/*T* for the determination of the activation energy for the recovery of nickel with magnesium were 0 and 3 wt-%.

From the values of the kinetic parameters determined in this section, it can be stated that, as expected, the higher values of the rate constants indicated that the reaction rate was improved by the presence of magnesium in the alloys, at least initially. This occurred because, for reaction times greater than 70 min, the reaction tends to stop. According to the observations in the micrographs of the partially reacted particles, it can be stated that this occurred through the formation of a thin layer of reaction products around the NiO particles. In addition, the activation energy for the aluminothermic reduction reaction of the NiO particles when using alloys with 3 wt-% by weight of magnesium, was lower than that of the experiments when no Mg was added.

3.4. Mechanism of Reaction

Regarding the reaction mechanisms that operate during the aluminothermic reduction reaction of NiO powders, the most precise explanation is that given by Zhong et al. [31], who stated that the formation of $MgAl_2O_4$ depends strongly on the initial concentration of magnesium, as has been previously established thermodynamically and has been experimentally proven in this work. Therefore, for initial magnesium concentrations of around 1 wt-%, the stable phase is $MgAl_2O_4$. Mcleod and Gabryel [32] established that the presence of MgO occurred because of the high initial concentration of magnesium (>3 wt-%). However, this concentration was easily reached at the NiO interface, especially in the initial stages after the addition of the particles, which ensured the formation of a significant amount of MgO. On the other hand, most of the MgO was consumed by the $MgAl_2O_4$ formation reaction given by Equation (4). Molins et al. [33] studied the interfacial reaction between a molten AlMg alloy and Al_2O_3 fibers, where it was shown that the MgO nuclei remained small and formed thin layers of ~10 μm thick. Therefore, the diffusion of magnesium occurred through the grain boundaries of the MgO particles by means of an infiltration mechanism. The growth process of the MgO layers continued until the grains around the matrix/reaction zone interface were large enough to close the intergranular spaces. The Gibbs free energy reaction of Equation (2) indicated that the reduction of NiO by aluminum occurred because Al can diffuse through the spaces left by the backwards diffusion of Ni. The diffusion of nickel occurred through the grain boundaries of MgO and Al_2O_3, dissolving in the molten aluminum when it reached the outer interface. Upon solidification, the nickel was rejected from the solid solution, forming Al_3Ni crystals around the partially reacted NiO particles. When the stoichiometric amounts of MgO and Al_2O_3 formed and equilibrium conditions occurred, the reaction given by Equation (2) was carried out. In this way, the formation of MgO and Al_2O_3 and the interdiffusion of Al, Mg, and Ni occurred simultaneously. After the magnesium concentration in the molten alloy fell below 1.98 wt-%, the aluminum continued to react with the NiO cores to form additional Al_2O_3 at a constant growth rate controlled by the diffusion of the aluminum through the layers of the reaction products. On the other hand, the small amount of MgO that could be formed by the magnesium reaction with NiO was dissolved in the Al_2O_3 phase. The reactions between the molten aluminum and magnesium dissolved with solid NiO require the diffusion of the atoms, although the diffusion of oxygen is negligible due to its large ionic size.

Therefore, the interdiffusion of magnesium and aluminum is kinetically possible through the network in the NiO structure of the O atoms, or through the nickel that results from the decomposition of the NiO. When the magnesium atoms occupy the interstitial sites within the NiO network, the position of the oxygen atoms must be adjusted to maintain electrical neutrality and to decrease the distortion energy of the network. Such reactions were given by the following:

$$Mg^0 - 2e^- = Mg^{+2} \quad (14)$$

$$2\left(Al^0 - 3e^0\right) = 2Al^{+3} \quad (15)$$

$$4\left(Ni^{+3} - 3e^-\right) = 4Ni^0 \quad (16)$$

The result is a change in the crystalline structure. The process involved in this phenomenon is governed by the values of the chemical potential of the partial reactions given, in this case, by Reactions (1) and (2). As $MgAl_2O_4$ is the final reaction product at room temperature, the reaction given by the global Equation (2) can be accepted, as this reaction satisfies the described mechanism. A necessary kinetic condition for the diffusion of chemical species is that the porosity remains in the layers of the reaction products. For similar reduction chemical reactions in molten aluminum, Zhong et al. [31] determined that the formation of MgO on SiO_2 particles involved a volume contraction of 13.6%, while the formation of $MgAl_2O_4$ on SiO_2 particles caused a volume contraction of 27%. Due to these changes in volume, the newly formed phases instantly broke their junctions with the original particles and transformed into thousands of small crystals. This transformation produces the necessary voids for the diffusion of the chemical species. However, for the last stages of the reduction process, the reduction mechanism changed abruptly, as the concentration of magnesium in the reaction interface decreased to below 1.98 wt-%. At this time, the MgO that was formed dissolved in Al_2O_3. Then, these Al_2O_3 crystals grew continuously until they reached a micrometric size, which resulted in the blocking of spaces for diffusion and the chemical reaction begins to stop. In this last stage, the efficiency of the reaction decreased sharply. The above explanations are important from the technological point of view, as unreacted NiO particles can be trapped in the molten metal as inclusions, and thus also affect the efficiency of the reaction. Of course, the kinetics of the reaction can be accelerated by further decreasing the size of the NiO particles or imposing the mixing conditions in the turbulent flow regime by using Reynolds numbers above 4.5×10^3 to break the layers of the reaction products once they have formed [34].

Figure 18 shows a schematic representation of the proposed mechanism of reaction, which described the different stages and the reactions among the participant chemical species, corresponding to a situation where the temperature was constant at 1123 K, initial magnesium concentration was constant at 3 wt-%, and agitation speed is constant at 300 rpm.

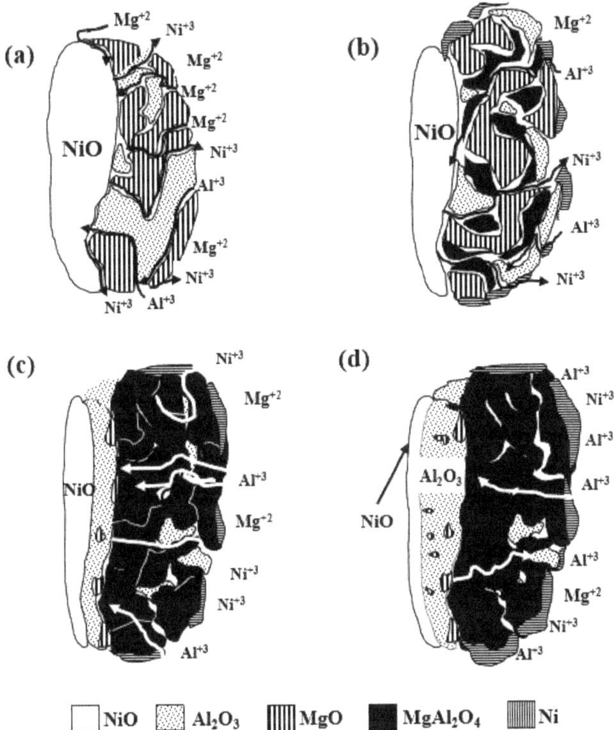

Figure 18. Scheme of the mechanism of reaction proposed for the permanent contact reaction between dissolved magnesium in molten aluminum and NiiO particles; (**a**) instantaneous formation of MgO and Al_2O_3, 10 min; (**b**) $MgAl_2O_4$ formation, 40 min; (**c**) nickel diffusion to the molten alloy, 70 min; (**d**) and consolidation of the layers of reaction products consisting of $MgAl_2O_4$ separated by layers of the Al_2O_3 phase, 90 min.

4. Conclusions

1. The aluminothermic reduction of NiO was studied at a laboratory scale by means of the powder injection technique, assisted by mechanical agitation, and achieved an increase in the nickel concentration in the solidified alloys of up to 3 wt-% for some of the experiments carried out.
2. It was observed that increasing the temperature favored the increase in the nickel concentration, because the mechanisms that govern the kinetics of the process, diffusion to the boundary layer, diffusion inside the layers of reaction products, and chemical reaction, were thermally activated. Increasing the mixing conditions by increasing the stirring speed of the molten bath using mechanical agitators at the velocity of 300 rpm promoted a greater agitation, therefore improving the efficiency of the reaction.
3. Increasing the initial magnesium concentration in the molten alloy allowed the Ni concentration to increase in the alloys, as was thermodynamically and experimentally demonstrated.
4. The experimentally obtained values of the Ni concentration as a function of time for different values of the parameters investigated (temperature, agitation speed, or initial Mg concentration) were adjusted to the kinetic equation of the diffusion model, which adjusted reasonably well. This allowed us to determine the values of some kinetic parameters of interest. On the other hand, the activation energy of the processes based on the Arrhenius equation was determined to be 15.80 KJ mol^{-1} for alloys containing an initial amount of 3 wt-% Mg.

5. With respect to the reaction mechanism, it was found that the step that controlled the overall chemical reaction was the diffusion of the Al and Mg atoms to the boundary layer, where they reacted with NiO particles, releasing Ni and forming Al_2O_3 and MgO as the reaction products. In turn, these compounds formed $MgAl_2O_4$ during cooling. The formation and breaking of $MgAl_2O_4$ into many crystals ensured the porosity required for the diffusion of the chemical species involved.

Author Contributions: Investigation, C.S.B. and R.O.P.; supervision, A.F.V. and J.T.T.

Funding: This research was funded by Fundicion J.V. with the project 217843 of the program Stimulus to the Investigation of CONACYT Mexico.

Acknowledgments: The authors wish to acknowledge the financial support of Fundicion J.V., project 217843 of the program Stimulus to the Investigation of CONACYT México, as well as for the cast shop facilities provided through this project.

Conflicts of Interest: The authors declare no conflicts of interest.

References

1. López, L.N.; Fernandes, H.; Gutiérrez, A.; Sánchez, J.A. Turning of thick thermal spray coatings. *J. Therm. Spray Technol.* **2001**, *10*, 249–254.
2. Beranoagirre, A.; Lopez de Lacalle, L.N. Optimising the milling of titanium aluminide alloys. *Int. J. Math. Math. Sci.* **2010**, *3*, 425–436. [CrossRef]
3. Priarone, P.C.; Klocke, F.; Giulia, M.; Lung, D.; Settineri1, L. Tool life and surface integrity when turning titanium aluminides with PCD tools under conventional wet cutting and cryogenic cooling. *Int. J. Adv. Manuf. Technol.* **2016**, *85*, 807–816. [CrossRef]
4. Beranoagirre, A.; Olvera, D.; López de Lacalle, L.N. Milling of gamma titanium–aluminum alloys. *Int. J. Adv. Manuf. Technol.* **2012**, *62*, 83–88. [CrossRef]
5. Quintana, R.; Perdomo, L.; Cruz, A.; Gomez, L.; Garcia, L.L.; Cerpa, A.; Cores, A. Obtención simultanea de ferroaleación multicomponente y escoria a partir de arenas negras, para el desarrollo de consumibles de soldadura por arco eléctrico. *Rev. Metal.* **2004**, *40*, 294–303. [CrossRef]
6. Ai, D.; Liu, K.; Lu, Z.; Zou, M.; Zeng, D.; Jun, M. Aluminothermal synthesis and characterization Li_3V_2-xAlx$(PO_4)_3$ cathode materials for lithium ion batteries. *Electrochim. Acta* **2011**, *56*, 2823–2827. [CrossRef]
7. Muñiz, R.; Flores, A.; Torres, J. A kinetic study of the strontium extraction by metallothermic reduction using submerged SrO powders injection. *Mater. Lett.* **2008**, *62*, 637–640. [CrossRef]
8. Luna, S.; Flores, A.; Muñiz, R.; Fernández, A.; Torres, J.; Rodriguez, N.; Ortíz, J.C.; Orozco, P. Cerium extraction by metallothermic reduction using cerium oxide powder injection. *J. Rare Earths* **2011**, *29*, 74–77. [CrossRef]
9. Flores, A.; Juárez, R.; Torrres, J.; Ayala, Z. A kinetic study on the aluminothermic reduction of ZrO_2. *J. Eng. Technol.* **2012**, *2*, 17–22.
10. Guía, J.C. Estudio de la Reducción Metalotérmica de Una Mezcla de Fe_3O_4-Cr_2O_3 para la Obtención de Fe-C. Master's Thesis, CINVESTAV-Saltillo, Ramos Arizpe, Mexico, 2010.
11. Ochoa, R.M. Utilización de Electrodos de Baterías Alcalinas Descargadas como Materia Prima para la Elaboración de Aleaciones Al-Zn-Mg y Al-Mn'. Master's Thesis, CINVESTAV Saltillo, Ramos Arizpe, Mexico, 2009.
12. Udhayabanu, V.; Singh, N.; Murty, B.S. Mechanical activation of aluminothermic reduction of NiO by high energy ball milling. *J. Alloys Compd.* **2010**, *497*, 142–146. [CrossRef]
13. López, F.J. Estudio Termodinámico y Cinético de la Reducción de $SrCO_3$ por Al Bajo Condiciones de Vacío. Master's Thesis, CINVESTAV Saltillo, Ramos Arizpe, Mexico, 2006.
14. Muñiz, C.R. Estudio del Proceso de Elaboración de Aleaciones Maestras Al–Si–Sr Mediante la Inyección de Polvos de $SrCO_3$. Ph.D. Thesis, CINVESTAV Saltillo, Ramos Arizpe, Mexico, 2008.
15. Zhao, K.; Feng, N.; Wang, Y. Fabrication of Ti-Al intermetallics by a two-stage aluminothermic reduction process using Na_2TiF_6. *Intermetallics* **2017**, *85*, 156–162. [CrossRef]

16. La, P.; Li, Z.; Li, C.; Hu, S.; Lu, X.; Wei, Y.; Wei, F. Effect of substrates on microstructure and mechanical properties of nano-eutectic 1080 steel produced by aluminothermic reaction. *Mater. Charact.* **2014**, *92*, 84–90. [CrossRef]
17. Mishra, K.; Zheng, J.; Patel, R.; Estevez, L.; Jia, H.; Luo, L.; Khoury, P.E.; Li, X.; Zhou, X.D.; Zhang, J.G. High performance Si@C anodes synthesized by low temperature aluminothermic reaction. *Electrochim. Acta* **2018**, *269*, 509–516. [CrossRef]
18. Liu, F.G.; Wen, C.D.; Hu, X.W.; Gao, B.L.; Shi, Z.N.; Wang, Z.W. Preparation of aluminum-zirconium master alloy by aluminothermic reduction in cryolite melt. *JOM* **2017**, *69*, 2644–2647. [CrossRef]
19. Malekia, A.; Hosseini, N.; Niroumand, B. A review on aluminothermic reaction of Al/ZnO system. *Ceram. Int.* **2018**, *44*, 10–23. [CrossRef]
20. Ochoa, R.; Flores, A.; Torres, J. Effect of magnesium on the aluminothermic reduction rate of zinc oxide obtained from spent alkaline battery anodes for the preparation of Al-Z-Mg alloys. *J. Int. Miner. Metall. Mater.* **2016**, *23*, 458–465. [CrossRef]
21. HSC Chemistry Software, Version 6.12. Outotec. 2007. Available online: https://www.outotec.com (accessed on 9 July 2018).
22. Thermochemical Software and Data Base System, Version 6.1. FactSage. 2009. Available online: http://www.factsage.com/ (accessed on 9 July 2018).
23. Guedes, M.; Ferreira, J.M.; Ferro, A.C. A study on CuO-Al_2O_3 infiltration by aluminium. *Mater. Sci. Forum* **2010**, *636–637*, 571–577. [CrossRef]
24. Langlais, J.; Harris, R. Strontium extraction by aluminothermic reduction. *Can. Metall. Q* **1991**, *31*, 127–131. [CrossRef]
25. Pai, B.C.; Ray, S. Fabrication of aluminum–alumina (magnesia) particulate composites in foundries using magnesium additions to melts. *Mater. Sci. Eng. A* **1976**, *24*, 31–44. [CrossRef]
26. Gul, F.; Karakulak, E.; Yamanoglu, R.; Zeren, M. Mechanical properties of Al-Ni Cast alloys. In Proceedings of the 23rd International Conference on Metallurgy and Materials, Brno, Czech Republic, 21–23 May 2014; pp. 1283–1287.
27. Bergsmark, E.; Simensen, C.J.; Kofstad, P. The oxidation of molten aluminum. *Mater. Sci. Eng. A* **1989**, *120–121*, 91–95. [CrossRef]
28. Ochoa, R. Relación Microestructura—Propiedades Mecánicas de las Aleaciones Al-Zn-Mg-Cu y Al-Zn-Mg-Cu-1% Li Elaboradas por Reducción Aluminotérmica del Ánodo de Pilas Alcalinas Descargadas y Latas para Bebidas. Ph.D. Thesis, CINVESTAV Saltillo, Ramos Arizpe, Mexico, 2016.
29. Allaire, C. Furnaces: Improving low cement castable by non-wetting additives. *JOM* **2001**, *53*, 24–27.
30. Brown, M.E.; Gallagher, P.K. *Handbook of Thermal Analysis and Calorimetry*; Elsevier: London, UK, 2008; pp. 148–149.
31. Zhong, W.M.; L'Espérance, G.L.; Suery, M. Interfacial reactions in Al-Mg (5083)/SiC*p* composites during fabrication and remelting. *Metall. Mater. Trans. A* **1995**, *26*, 2625–2635. [CrossRef]
32. Mcleod, A.D.; Gabryel, C.M. Kinetics of the growth of spinel, $MgAl_2O_4$, on alumina particulate in aluminum alloys containing magnesium. *Metall. Mater. Trans. A* **1992**, *23*, 1279–1283. [CrossRef]
33. Molins, R.; Bartout, J.D.; Bienvenu, Y. Microstructural and analytical characterization of Al_2O_3-(Al-Mg) composite interfaces. *Mater. Sci. Eng. A* **1991**, *135*, 111–117. [CrossRef]
34. Dimotakis, P.E. The mixing transition in turbulent flows. *J. Fluid. Mech.* **2000**, *409*, 69–98. [CrossRef]

Article

Uranium Removal from Groundwater by Permeable Reactive Barrier with Zero-Valent Iron and Organic Carbon Mixtures: Laboratory and Field Studies

Borys Kornilovych [1,*], Mike Wireman [2], Stefano Ubaldini [3], Daniela Guglietta [3], Yuriy Koshik [4], Brian Caruso [5] and Iryna Kovalchuk [6]

1 National Technical University of Ukraine "Igor Sikorsky Kyiv Polytechnic Institute", 37 Peremogy Av., 03056 Kyiv, Ukraine

2 US Environmental Protection Agency—Granite Ridge Groundwater, Boulder, CO 80305, USA; graniteridgegw@q.com

3 Istituto di Geologia Ambientale e Geoingegneria, CNR, Area della Ricerca di Roma RM 1—Montelibretti-Via Salaria Km 29,300—Monterotondo Stazione, 00015 Roma, Italy; stefano.ubaldini@igag.cnr.it (S.U); daniela.guglietta@igag.cnr.it (D.G.)

4 Ukrainian Scientific Research and Design Institute for Industrial Technology, Ministry of Energy and Coal Industry of Ukraine, 37 Petrovsky Str., 52220 Zhovty Vody, Dnipropetrovsk Region, Ukraine; ipt@iptzw.org.ua

5 US Environmental Protection Agency: Currently US Fish and Wildlife Service, Department of the Interior, 134 Union Blvd, Lakewood, CO 80228, USA; brian_caruso@fws.gov

6 Institute for Sorption and Problems of Endoecology, National Academy of Science of Ukraine, 13 General Naumov Str., 03164 Kyiv, Ukraine; kowalchukiryna@gmail.com

* Correspondence: b.kornilovych@gmail.com; Tel.: +38-050-840-7970

Received: 31 March 2018; Accepted: 24 May 2018; Published: 1 June 2018

Abstract: Zhovty Vody city, located in south-central Ukraine, has long been an important center for the Ukrainian uranium and iron industries. Uranium and iron mining and processing activities during the Cold War resulted in poorly managed sources of radionuclides and heavy metals. Widespread groundwater and surface water contamination has occurred, which creates a significant risk to drinking water supplies. Hydrogeologic and geochemical conditions near large uranium mine tailings storage facility (TSF) were characterized to provide data to locate, design and install a permeable reactive barrier (PRB) to treat groundwater contaminated by leachate infiltrating from the TSF. The effectiveness of three different permeable reactive materials was investigated: zero-valent iron (ZVI) for reduction, sorption, and precipitation of redox-sensitive oxyanions; phosphate material to transform dissolved metals to less soluble phases; and organic carbon substrates to promote bioremediation processes. Batch and column experiments with Zhovty Vody site groundwater were conducted to evaluate reactivity of the materials. Reaction rates, residence time and comparison with site-specific clean-up standards were determined. Results of the study demonstrate the effectiveness of the use of the PRB for ground water protection near uranium mine TSF. The greatest decrease was obtained using ZVI-based reactive media and the combined media of ZVI/phosphate/organic carbon combinations.

Keywords: uranium; contaminated groundwater; permeable reactive barrier; zero-valent iron

---

1. Introduction

The uranium industry is critically important for the Ukrainian economy: nearly 45% of all electric power is produced by nuclear power stations. Ukraine produces more uranium than any other European country and in 2016 was the 10th largest producer in the world: production of uranium from mines in Ukraine was 1006 t/year [1]. The largest uranium deposits occur in Kryvyi Rig basin located in the western part of Dnipropetrovsk Oblast (Figure 1) and are typically associated with iron ore deposits.

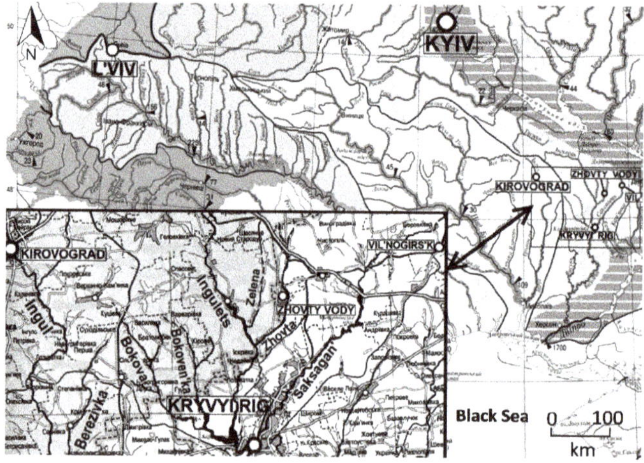

Figure 1. Kryvyi Rig Basin location.

Significant environmental pollution in the Kryvyi Rig basin relates to industrial activity in Zhovty Vody city. The largest is the state enterprise for uranium ore mining and milling—"Eastern Mining and Concentration Combine Works" (SE "VostGOK"). Waste products associated with uranium processing represent a major source of surface and groundwater contamination [2,3].

Throughout the United States and Western Europe numerous methods have been utilized for remediating water, soil and sediments contaminated with hazardous substances. A relatively new (in Eastern Europe) and potentially important in situ method is a permeable reactive barrier (PRB) constructed in the subsurface to intercept and treat contaminated groundwater which contain dissolved metals, radionuclides, and dissolved nutrients. During recent years, the use of PRBs has evolved from innovative to accepted standard practice for treatment of selected organic and inorganic contaminants. There have been more than 200 successful PRB applications all over the world, but mainly in the United States, and the US Environmental Protection Agency (EPA) designated the PRB a standard remediation technology [4–7].

The contaminant treatment zone in a PRB may be created directly using reactive materials such as iron or indirectly using materials designed to stimulate secondary physical, chemical, or biological processes. Various materials have been investigated as possible treatment materials for use in a PRB, including zeolite, hydroxyapatite, elemental iron, limestone, and others [7–11].

One of the most promising treatment materials for use in a PRB is zero-valent iron (Fe^0) which has been extensively studied in recent years [12–17]. It has been shown that iron fillings are much more effective than adsorbents in removing heavy metals and uranyl (UO_2^{2+}) from the aqueous solution. However, it is still unclear whether the removal of heavy metals and uranyl results from reductive precipitation or from adsorption onto the corrosion products of Fe^0. A more exact knowledge of the chemical interactions in PRB treatment zones is needed to improve the efficiency of new designs and to predict longevity so that reliable cost assessments can be made [18–20].

It is necessary to note that in some cases of PRB applications for groundwater remediation have failed to achieve cleanup levels as predicted from bench scale tests, for molybdenum and uranium [21]. Performance failure was related to a sequence of events due to continual buildup of mineral precipitates on the reactive medium, loss of pore space, development of preferential flow paths, and finally complete bypass of the zero-valent iron resulting in the loss of hydraulic control.

The biological approach in PRB technology is also an innovative method that can be useful for remediating uranium contamination of water. Many environmental biotechnological applications utilize microorganisms that have key roles in the biogeochemical cycling of toxic metals and radionuclides [22–27]. Biologically induced reduction reactions can promote the attenuation of inorganic cations through direct reduction of the metal or through indirect precipitation resulting from the oxidation or reduction of an inorganic anion. Many metals, including toxic metals such as Cd, Cu, Zn, Mn, Pb are removed by this method [28–30].

In this study, special attention was devoted to uranium. It was shown that microbes had the ability to facilitate the removal of uranium from water through the sorption of U(VI) to bacterial cell walls, biological reduction of U(VI) and enzymatic production or nucleation of U mineral precipitates [25,31–33]. Dissolved oxygen is reduced first, followed by denitrification, followed by UO_2CO_3 reduction to uraninite and sulfate reduction to sulfide [25]. As a result, an amorphous uranium oxide may form on bacteria surfaces [34].

Biological reduction of U(VI) to insoluble U(IV) was stimulated by the addition of ethanol and trimetaphosphate to contaminated groundwater [27,35]. Also, kinetic data for microbial U(VI) reduction were obtained [36]. The process in which *Citrobacter* sp. produces an overabundance of alkaline phosphatase was described, which in turn causes the cell to excrete phosphate, which provides the nucleus for the precipitation of U(VI) phosphate on cell surfaces [37].

The role of inorganic support in microbiologically-induced reduction of U(VI) is very important [38]. Another very perspective approach in PRB application is to combine chemical oxidation-reduction reactions and microbiologically mediated processes [39,40].

The main objectives of this work were (1) to characterize the site at Ukrainian Uranium Center (Zhovty Vody city, Ukraine) with respect to the design and installation of a PRB to treat contaminated groundwater and (2) to study the effectiveness of different zero-valent iron and organic carbon fillings for uranium removal by PRB technology.

2. Materials and Methods

2.1. Water Sampling and Analysis

Zhovty Vody and the uranium mining district are located within the Kryvyi Rig iron-ore basin. This structurally complex basin (500–600 km^2) occurs within the central part of the Ukrainian Shield and is underlain by Precambrian volcanic, granitic, and meta-sedimentary rocks overlain by thin, unevenly distributed Cenozoic sedimentary rocks. A variety of Quaternary alluvial, eolian and lacustrine deposits form the surface over most of the basin. The uranium deposits occur in hydrothermally altered rocks, are associated with the iron deposits, and have strata-like and lens-like forms. The deposits are typically 1–15 m thick, tens to hundreds of meters long and occur at depths up to 1.5 km.

Since 1950s, SE "VostGOK" performed uranium ore underground mining and processing by hydrometallurgical method. Currently, hydrometallurgical plant processes ores being delivered from Ukrainian uranium deposits and produces natural uranium concentrate (uranous-uranic oxide—U_3O_8). The technological process is based on leaching of the uranium from ore by sulfuric acid and sorption on anionic resins (so called "resin in pulp process"). Also, underground leaching and block leaching were used.

Numerous waste disposal facilities related to uranium mining and processing have been constructed near Zhovty Vody. Liquid and solid process wastes (disposal sludge, waste water from uranium ore processing and decontamination of process equipment, etc.) are discharged through slurry pipelines into tailings storage facilities. The tailings storage facility (TSF) "Shcherbakovskoyeh" ("Shch") (Figure 2) is located about 1.5 km south of Zhovty Vody in the Shcherbakovskaya gully which is a right tributary of the Zhovta River. For many years leachate from this facility has infiltrated into the sub-surface, contaminating the underlying groundwater.

Shcherbakovskaya gully and the Shch TSF are underlain by loams, clays, and sands of primarily Quaternary age. These surficial deposits overlie highly weathered Precambrian granite with a thick residual layer at the top which functions as an aquifer. Between the Shch TSF and the Zhovta River (1.0 km to the east) all but 4–5 m of the Quaternary sediments have been eroded. The highly weathered granite is quite permeable and has a significant saturated thickness. Groundwater in the granite flows eastward from the area of the Shch facility and discharges to the recent alluvial sediments along the Zhovta River and directly to the Zhovta River. The gradient on the water table ranges from 0.023–0.4. The Zhovta River flows southward and discharges to the Ingulets River south of the Zhovty Vody. The Ingulets River discharges to the Black Sea.

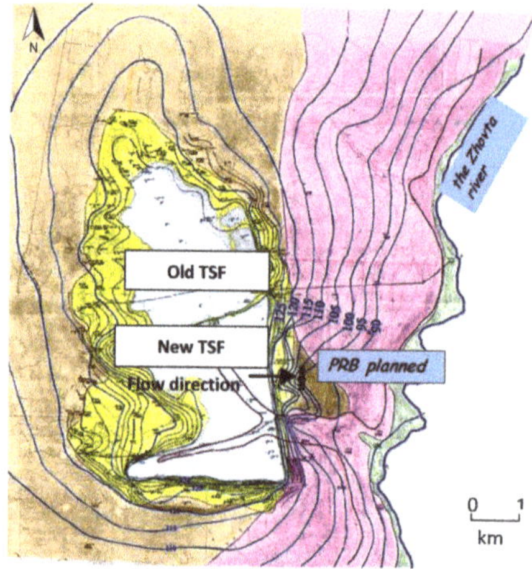

Figure 2. Hydrogeological map of the tailings storage facility site with hydroisohypses of the aquifer.

Currently, the TSF consists of two sections: an old tailings storage facility located in the northern branch of the Shcherbakovskaya gully and a new tailings storage facility located in the western branch of the gully (Figure 2). The old TSF was constructed by building a dam across the northern branch of the gully and is 1.8 km long, 0.9 km wide, and 98.4 ha in area. Its storage volume is about 5.4×10^6 m^3. It was operated from 1959 to 1979 and is currently used as a standby facility. The new TSF, in operation since 1979, has an area of 151.8 ha, and a projected volume of 25.84×10^6 m^3. Solid and liquid process wastes from the hydrometallurgical plant include pulp from uranium leaching, waste water from washing and decontamination of process equipment etc. Sludge drainage and industrial sewage from the neutralization station from the sulfuric acid plant are also discharged through slurry pipelines into the tailings storage facility. The waste is emplaced under the layer of water to prevent its drying and dusting.

The chemistry of the groundwater in the weathered granite aquifer has been highly contaminated by leachate infiltrating from the Shch TSF. Data from sampling near the facility indicate a groundwater contamination plume defined by a high concentration halo of nitrate and sulfate. The configuration of the plume corresponds well with the direction of groundwater flow. Dissolved solids concentrations in groundwater to the west (upgradient) of the tailings storage facility ranged from 1.0 to 1.95 g/L. Downgradient of the tailings facility the dissolved solids concentrations ranged from 2.12 to 5.33 g/L. Groundwater is mainly a sulfate, magnesium-sodium type. The pH ranged from 6.8 to 8.0 and the total hardness ranged from 12.04 to 86.97 mg-equiv/L.

These data, obtained from a network of the monitoring wells installed by the enterprise SE "VostGOK", clearly indicate the presence of leachate from the tailings storage facility in the groundwater. Groundwater near the TSF was sampled from 8 monitoring holes to determine chemical and radiochemical (radionuclides of uranium series) composition. Detailed geochemical sampling was conducted in December of 2011, as well as March, May, August, November 2012 and April, June 2013. Sampling procedure was consistent with international standards, such as established by the International Organization for Standardization and the US Environmental Protection Agency [41,42].

Position of the groundwater levels was measured in all monitoring holes characterizing efficiency of the permeable reactive barrier using hydrogeological roulette with periodicity 2 times a month. Groundwater was collected using a sinking pump. Water samples were collected in disposable 0.06 L syringes and filtered through 0.45 μm cellulose acetate filter. Ultrapure concentrated HNO_3 was added (pH < 2) to preserve the cation sample. All samples were analyzed at three different chemistry and radiochemistry laboratories: at National Technical University of Ukraine (Kyiv), Ukrainian Research-Development and Design-Prospecting Institute of Industrial Technology (Zhovty Vody) and East Mine Dressing Combine (Zhovty Vody). The concentration of metal ions (Na^+, K^+, Ni^{2+}, Cu^{2+}, Zn^{2+}, Pb^{2+}, Cd^{2+}, Fe_{tot}, Co_{tot}, Mn_{tot}) was determined by atomic absorption spectroscopy (AAS) with AA-6300 Shimadzu. Other chemical constituents (Mg^{2+}, Ca^{2+}, NH_4^+, HCO_3^-, Cl^-, SO_4^{2-}, NO_3^-, total dissolved solids (TDS)) were determined by standard chemical procedures [43]. Dissolved uranium (in this and next sections) was quantified spectrophotometrically on UNICO 2100UV using the Arsenazo III (1,8-dihydroxynaphthalene 3,6-disulphonic acid-2,7-bisSTA(azo-2)-phenylarsonic acid) reagent. The chemical composition of this groundwater is shown in Table 1.

Table 1. Chemical composition of contaminated groundwater from locations near planned PRB (mg/L, pH 7.2).

Ca^{2+}	Mg^{2+}	$Na^+ + K^+$	NH_4^+	HCO_3^-	Cl^-	SO_4^{2-}	NO_3^-	TDS
576	209	391	0.92	448	182	2832	125	4007
U_{sum}	Ni^{2+}	Cu^{2+}	Co^{2+}	Mn_{sum}	Zn^{2+}	Pb^{2+}	Cd^{2+}	Fe_{sum}
0.42	<0.05	<0.03	<0.06	0.10	<0.01	<0.19	<0.01	0.05

The data indicate that the concentrations of uranium, SO_4^{2-}, and total dissolved solids (TDS) significantly exceed international standards. The maximum contaminant levels established by the US Environmental Protection Agency are: 0.03 mg/L for uranium, 250 mg/L for SO_4^{2-} and 500 mg/L for TDS [44].

2.2. Experimental System

Two main approaches were applied in laboratory experiments with PRB treatment media: chemical and microbiological. The chemical tests can be performed under no flow conditions (batch tests) or under gravitational or imposed flow conditions (column tests).

The first step in the chemical approach was conducting batch tests to aid in the preliminary selection of possible reagents and to help characterize the effect of contaminated groundwater (constituents) on the reactivity of Fe^0.

For this purpose, batch experiments were conducted with three different types of iron materials (all from Ukrainian Plants). The first one, zero-valent iron material 1 (sample: ZVI1) was a direct reduced iron (Fe > 99.5%) with particle size < 0.16 mm. The second one (sample: ZVI2) was iron material (Fe > 99.0%) with particle sizes as follows: >1.0 mm—0.4%; 1.0–0.63 mm—21.7%; 0.63–0.4 mm—16.9%; 0.4–0.31 mm—10.5%; 0.31–0.2 mm—19.9%; 0.2–0.16 mm—13.9%; 0.16–0.1 mm 11.6%; 0.1–0.063 mm—1.8%; 0.063–0.05 mm—1.0%; <0.05 mm—2.3%. The third material (sample: ZVI3) was scrap iron (C—3.3%; Si—2.0%; Mn—0.8%) with particle sizes between 1.0 and 2.0 mm.

These experiments consisted of placing a known mass of the Fe-powder in glass (or non-reactive) flasks, filling them with the contaminant solution and shaking at various times. At selected times samples of solutions from all flasks were collected and analyzed to determine uranium concentration changes and obtain physical-chemical parameters (e.g., pH, electric conductivity, etc.).

The soluble part of humic substances, fulvic acid (FA), and Na_2CO_3 were used in the experiments to determine removal efficiency of iron materials for uranium solutions that contain complexing agents. These substances are common constituents in ground water and readily form complexes with U(VI) ions [45,46]. FA was extracted from lake sediments near Kyiv (Ukraine) according to literature method [43]. Ethylenediaminetetraacetic acid (EDTA), nitrilotriacetic acid (NTA) and ethylenediamineacetic acid (EDA), which are common constituents in washing solutions [47], were also used. ZVI2 as zero-valent iron material was used in these experiments.

The second step in the chemical approach was a column test which was most important for designing the PRB. The test included percolation of a solution, with a known concentration of contaminant, through a cylindrical column filled with reagent material. Because all experiments were conducted using groundwater samples with very low concentration of dissolved oxygen, some precautions were taken to prevent contact of the solutions with air.

In the microbiological approach the same laboratory experimental material was used for testing PRB design and effectiveness. Iron, sand, gravel, bone meal and sawdust, which widely used as fillings for biological PRB [4,5], were selected for testing. An activator was also added to initiate bioreduction processes. It contained anaerobic bacterial communities with uranium-removal bacteria which were prepared from a previously cultivated mixed indigenous culture of bacteria from bottom sediments of Zhovta River near mine tailings storage. Yeast decoction and cysteine were added to the solution for biostimulation.

Various mixtures of materials were tested for further application as an active PRB treatment medium (detailed composition of the medium is in Section 2.3):

1. Powdered ZVI + sand (inorganic active filling);
2. Powdered ZVI + gravel + bone meal + sawdust + activator (organic-inorganic active filling);
3. Gravel + bone meal + sawdust + activator (organic active filling).

A sketch of the apparatus, which was designed and commissioned for testing PRB treatment media, is shown in Figure 3.

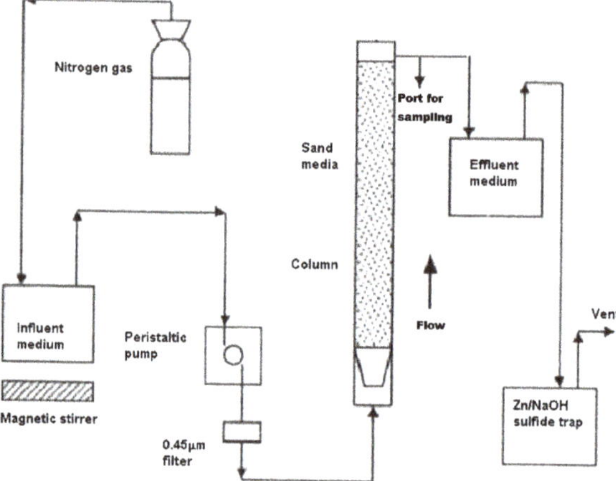

Figure 3. A schematic diagram of the up flow anaerobic packed bed reactor setup.

A packed bed system was constructed from Plexiglas with an overall height of 30 cm, internal diameter of 8 cm and a net empty working volume of 1 L. It was equipped with an 0.8 cm diameter port used for sampling liquid located 20 cm from the bottom. The flow was dispersed with the aid of a frustum shaped cowling located at the base of the column, which also served to contain the porous media.

The influent was pumped from the reservoir tank to the inlet of the reactor by means of a pre-calibrated variable speed peristaltic pump at a flow rate of 0.006–0.15 L/h (approximately 0.5–10 pore volumes per day). Inert gas (N_2 or CO_2) is continuously purged through the influent reservoir to lower the oxygen content of the influent solution and satisfy the anaerobic conditions during column experiments. A series of preliminary experiments were conducted to determine the operability of column apparatus. For these preliminary experiments inorganic materials were used as PRB active treatment medium. The stability of the equipment was demonstrated for all preliminary tests.

The effluent volume was recorded daily by weighing the mass of water collected, allowing the average flow rate during the collection period to be calculated. Columns were operated in different experimental series, but the total treated pore volumes for the individual columns differed during equal periods of time because of differences in the voidage.

2.3. Chemical and Microbial Tests

In the first series of column experiments sieved sand (particle size 0.6–2 mm) and iron material (ZVI2) were used. The ratio of the components in reactive media was 1:10 (Fe^0 powder to sand). After filling with the reactive mixtures, the columns were purged with several pore volumes of deoxygenated water to ensure complete removal of air (oxygen) prior to addition of the test solution.

The total duration of each series of experiments was ~90 days. The samples of purified water after columns were collected and analyzed with the use of special port for sampling every day during first month of the set and every 2–3 days during next two months. Calculation of the removed uranium was based on the difference between the influent and effluent concentrations. After the end of the experiment the leaching procedure was provided for determination of the uranium distribution in reactive media in the column. The whole reactive material from every column was divided into layers (1 cm) and then treated with HNO_3. The concentration of uranium in leaching solution from each layer was determined and then the total content was calculated. Also, it was interesting to determine the distribution of released uranium within the separate layers of the reactive media. The samples were taken at different points of separate reactive media layers and then analyzed.

An organic-inorganic mixture was used in a second series of anaerobic bioremediation experiments with an indigenous consortium of microorganisms from mine water. An organic component consisting of sewage sludge, bone meal, sawdust and activator was used with an inorganic component consisting of gravel and Fe^0 (ZVI2). All materials (except ZVI2 which was taking from enterprise "Public Plant of Powder Metallurgy", Brovary city, Ukraine) were obtained from local enterprises in Zhovty Vody. A mixed indigenous culture of bacteria as one of the components of the activator was also added to this composition. The ratio of Fe^0 powder, sewage sludge, bone meal, sawdust, water (with activator) and gravel in the mixture was chosen 0.17:1:2:3:4:17, in accordance with composition of typical reactive mixtures for biological PRBs [4,5,48].

The activator was prepared using sludge samples from bottom sediments of Zhovta River near the "Sch" mine tailings storage facility. A culture of bacteria for these tests was cultivated by suspending 0.04 L of the sediment in 0.4 L of basal medium in the presence of 50 mg/L U(VI). The propagation was continued with basal medium which was replaced weekly. Basal medium was modified from Postgate C medium with 4.5 g/L Na_2SO_4 as electron acceptor [49]. Resazurin (1 mg/L) was added as a redox indicator to show any contamination by molecular oxygen. The pH was adjusted to 7.5 with 50% NaOH (*w/v*). After four weeks of domestication, sulfate-reducing bacteria (SRB) for lab experiments were obtained. The activation was done periodically to maintain the activity of SRB. The columns were wet packed with all components. The experimental procedure was the same as in experiments with inorganic treatment medium.

In a third series of experiments only treatment medium of organic composition was used. Sewage sludge, bone meal, sawdust, water solution of activator and inert inorganic component (gravel) were mixed in the following ratio 1:2:3:4:17. A mixed indigenous culture of bacteria as one of the components of the activator was also used in this series of experiments.

Contaminated groundwater from sites near the uranium mine TSF was used in these column experiments. The concentration of uranium in influent was increased to 10 mg/L to approach the exhausting of the sorption capacity of the medium. This solution was pumped to the inlet of the column. All U(VI) reduction experiments were conducted anaerobically in the dark.

2.4. Field Tests

Sampling for radionuclide composition of groundwater in 8 monitoring holes in the PRB area was conducted to evaluate the efficiency of the PRB cylinders with the different types of reactive media.

Because of the significant risk the TSF poses to the Zhovta River and the favorable hydrogeological setting, the Shch tailings facility site was chosen for a PRB installation. The near surface geology at the PRB locations, as determined from numerous boreholes, is as follows:

- Diluvial soil organic layer of 0.9 m in thickness—the upper layer of soil;
- Black alluvial loam of 2.6 m in thickness—the middle layer of soil;
- Residual soil of crystalline rocks—gruss-gravel formations of 2.5 m in thickness—the lower layer of soil.

The crystalline rocks of Precambrian occur from a depth of 6.0 m. Results from our groundwater flow modeling, using VISUAL MODFLOW v. 2.8.2 (Waterloo Hydrogeologic Inc., 2000, Waterloo, ON, Canada) indicate that 93.8% of the groundwater discharge to the Zhovta River is from weathered, residual crystalline rocks.

The position of the PRB in the region of TSF is presented in Figure 2. The PRB consists of 21 cylinders installed in three clusters of boreholes (Figure 4). Each cluster of cylinders contained a different treatment media. The first cluster of boreholes (group A) contained inorganic active filling—Fe^0 powder and sand (T-1–T-7), second cluster (group B) contained organic-inorganic active filling—sewage sludge, bone meal, sawdust and water (with components of activator) and gravel (T-8–T-14) and third cluster (group C) contained organic active filling—sewage sludge, bone meal, sawdust, water solution of activator and gravel (T-15–T-21). The holes that house the treatment cylinders are 0.35 m in diameter. The holes in each cluster are staggered to make sure all groundwater flow is intercepted. The depth of the holes is 6 m.

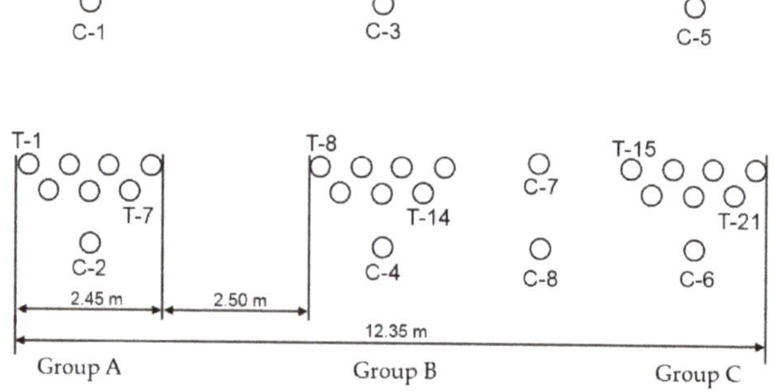

Figure 4. Scheme of PRB holes. Group A—cylinders filled with inorganic active filling; Group B—with organic-inorganic active filling; Group C—with organic active filling.

Sampling for radionuclide composition of groundwater in 8 monitoring holes in the PRB area (C-1–C-8) was conducted to evaluate the efficiency of PRB cylinders. Monitoring holes were installed up-gradient (holes C-1, C-3, C-5), within (C-7, C-8) and down-gradient of the PRB (C-2, C-4, C-6) (Figure 4).

3. Results and Discussion

3.1. Laboratory Tests

The results of the U(VI) removal by different types of ZVI are shown in Figure 5. The adjusted initial pH was 6. The greatest uranium removal occurred using samples of ZVI1 (the most dispersed iron material). The worst results were for scrap material ZVI3 with the largest particle size [50].

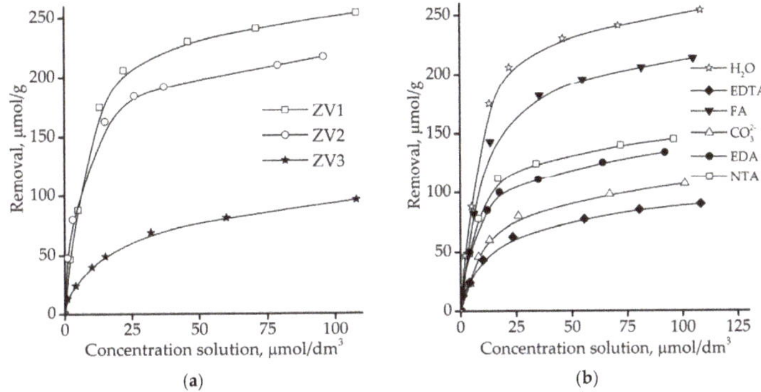

Figure 5. (**a**) Removal of U(VI) by different types of Fe^0; (**b**) Effect of selected ligands (FA, EDTA, NTA, EDA, and Na_2CO_3) on the U(VI) removal by ZVI1.

It was shown previously that reactivity of iron-containing materials may be considerably affected by some contaminants and groundwater constituents which demonstrate significant complexing of different metals [4,51].

Figure 5b compares the results of the U(VI) fixation by Fe^0 (ZVI1) for 3 days in the presence of 5×10^{-4} M EDTA, 5×10^{-4} M NTA, 5×10^{-4} M EDA, 10^{-3} M Na_2CO_3, and 100 mg/L FA solutions. These ligands readily form complexes with U(VI) ions and include the soluble part of humic substances, fulvic acid (FA), and CO_3^{2-}, which are common constituents in ground water; and ethylenediaminetetraacetic acid (EDTA), nitrilotriacetic acid (NTA) and ethylenediamineacetic acid (EDA), which are common constituents in washing solutions. Uranium removal efficiency diminished with the addition of all the ligands.

This is due to the changes in uranium speciation after the addition of carbonate and other complexing agents. One of the important uranium fixation mechanisms is sorption. As shown in the speciation diagrams in [52], positively charged uranyl (UO_2^{2+}) species dominate under CO_2-free conditions, while solutions in the presence of carbonate contain predominantly negatively charged $UO_2(CO_3)_2^{2-}$, $UO_2(CO_3)_3^{4-}$, or uncharged UO_2CO_3 complexes, with rather weak sorption properties. The same situation exists for other ligands [47,53]. Clearly, the lowest uranium removal from solution occurred with strong ligands such as EDTA and CO_3^{2-} and the highest with the weaker ligand FA.

In all the column tests almost complete U(VI) fixation occurred. Results (Figure 6a) show that practically full U(VI) removal was observed during all period of experiments (90 days).

The experiments showed that the distribution of retained uranium in the columns was not uniform—most of the uranium was concentrated in the lower layers of the reactive media near the bottom of the column (Figure 6b). Also, the specific contents of uranium at different points within the separate layers of the reactive media are not close by value, which indicates that the flows of the contaminated solution are not uniform in cross-section of the columns [54,55].

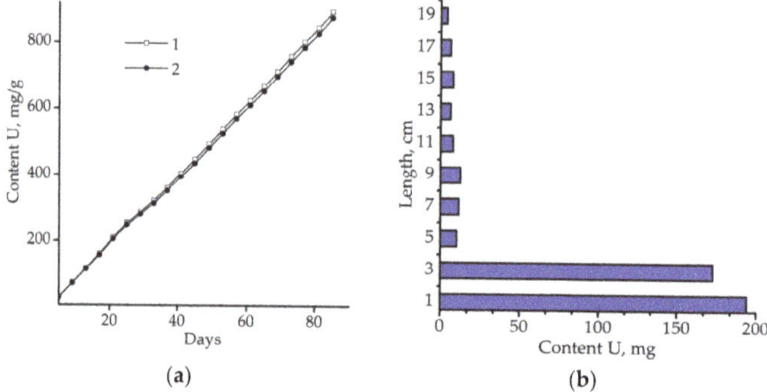

Figure 6. (**a**) Uranium released from contaminated water in column with ZVI2 and sand: 1—total uranium in the influent; 2—total released uranium; (**b**) Distribution of uranium in the reactive media in the column.

3.2. Field Tests

Water table contour maps for August 2012 and June 2013 are shown in Figure 7. These maps indicate a seasonal change in the direction groundwater flow. The dominant groundwater flow in the PRB area takes place towards the Zhovta River.

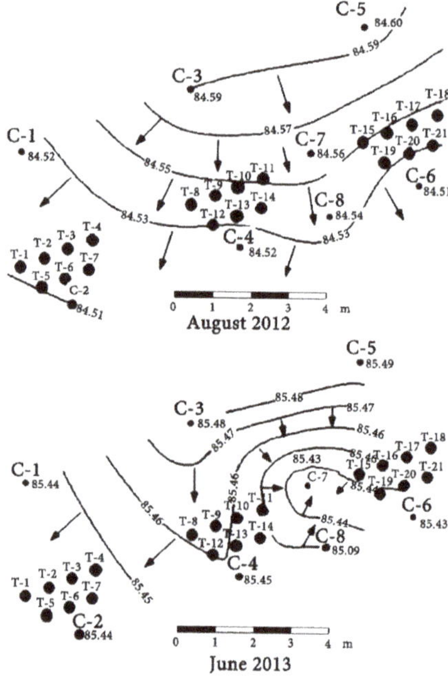

Figure 7. The change of groundwater flow direction (—▶) during the period of observation.

As stated previously, all (93.8%) groundwater flow to the Zhovta River occurs in the aquifer comprised of residual soil of crystalline rocks.

The position of 21 PRB cylinders and 8 monitoring holes on water table contour maps is presented in Figure 7.

Results for total uranium content (^{234}U, ^{238}U and ^{238}U) in groundwater from 8 monitoring holes are presented in Figure 8. The comparison of the obtained results for monitoring holes which were situated in pairs, before and after different groups of PRB cylinders, clearly indicate a considerable decrease in uranium contamination of groundwater after flow through the PRB with ZVI2.

The greatest decrease was observed in monitoring holes C-1 (Figure 7) and C-2, which are located near PRB cylinders filled with inorganic filling and in monitoring holes C-3 (Figure 7) and C-4 which are located near PRB cylinders filled with organic-inorganic filling. Smaller decreases were observed after flow through cylinders with organic filling (as indicated by data from monitoring holes C-5 (Figure 7) and C-6) because of the clayey and sandy character of the soil near the uranium mine TSF and the low content of microorganisms from the indigenous mine consortium.

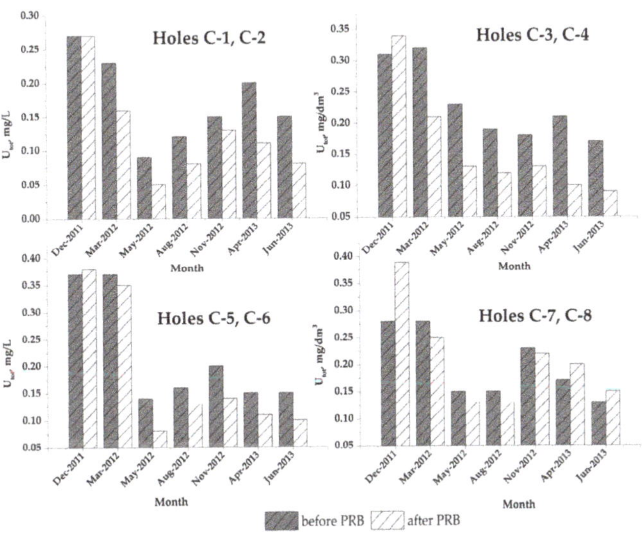

Figure 8. Uranium content in groundwater at PRB site: before PRB (monitoring hole C-1, C-3, C-5, C-7); after PRB (monitoring hole C-2, C-4, C 6, C-8) (Figure 7).

Even though groundwater flow is spatially and temporally complex at the PRB site the remedial performance of PRB is significant. The uranium content in groundwater flowing through the PRB was reduced by more than 50%. Therefore, a new design for a PRB is proposed. Such PRB construction with non-continuous reactive media allows reduction of the installation costs significantly. Even for the cheapest passive remediation methods, continuous wall and funnel-and-gate PRB systems, the costs of the PRB may be high depending of the construction length and especially the depth in the ground [4,5]. Instead of a conventional continuous barrier or a funnel-and-gate, this novel design consists of rows of cylinders with different reactive materials. This PRB design showed high efficiency for groundwater remediation of uranium contamination in the Ukrainian region of uranium mining and processing. The obtained results proved that the use of PRB with single and double rows of cylinders with active inorganic or inorganic-organic fillings provided necessary levels of groundwater purification in the case of low or middle uranium concentration (up to 0.3–0.4 mg/L). Three or more offset rows are needed for higher contaminant concentrations.

4. Conclusions

The results of laboratory tests of the U(VI) removal by different types of ZVI show that the greatest efficiency occurred using samples of the most dispersed iron material (zero-valent iron with very fine particle size—ZVI1). Uranium removal diminished with the addition of all most significant ligands presented in ground water.

A pilot-scale PRB was installed near the uranium tailing storage facility located in Ukrainian Uranium Center (Zhovty Vody city, Ukraine). The initial concentrations of uranium, SO_4^{2-}, and total dissolved solids in groundwater at that site significantly exceed international standards. A new design for a PRB is used. Instead of a conventional continuous barrier or a funnel-and-gate, this design consists of rows of cylinders with iron-reactive materials. Such PRB construction with non-continuous reactive media typically reduces allows considerable reduction of the installation costs.

After two years of monitoring a pilot-scale PRB, utilizing zero-valent iron and organic carbon treatment media, results indicate that uranium concentrations in groundwater at the PRB site were reduced from 0.38 mg/L to 0.07–0.15 mg/L. The greatest decrease was obtained using a treatment media of powder zero-valent iron and sand, and a treatment media of sewage sludge, bone meal, sawdust, water (with components of activator) and gravel.

Smaller decreases were observed using pure organic carbon treatment media because of the clayey and sandy character of the soil near the uranium mine TSF and the low content of microorganisms from the indigenous mine consortium. Probably the higher content of microorganisms (as in silt or black earth soil) stipulate the increasing of the level of groundwater purification.

Even in a complex groundwater flow system at the PRB site and the mine tailing storage facility the positive effect of PRB is significant, reducing the uranium content in groundwater at PRB site by 50%.

Author Contributions: Conceptualization, B.K., M.W. and B.C.; Methodology, B.K., M.W. and B.C.; Sampling and Analysis, Y.K. and D.G.; Field Tests, B.K. and Y.K.; Experiments, I.K. and D.G.; Data Analysis and Discussion, B.K., S.U. and I.K.; Writing-Original Draft Preparation, B.K., I.K. and S.U.

Funding: Funding for this research was supported by the U.S. Environmental Protection Agency, under projects P-322 "Regional evaluation of mining-related metals contamination, risks, and innovative remediation technologies in Ukraine and Georgia" and P-454 "Development of innovative environmental technology for remediation of contaminated groundwater in Ukraine" with the Science and Technology Center in Ukraine.

Conflicts of Interest: The authors declare no conflict of interest.

References

1. World Uranium Mining Production. Available online: http://www.world-nuclear.org/information-library/nuclear-fuel-cycle/mining-of-uranium/world-uranium-mining-production.aspx (accessed on 13 May 2018).
2. Kornilovych, B.; Wireman, M.; Caruso, B.; Koshik, Y.; Pavlenko, V.; Tobilko, V. The use of permeable reactive barrier against contaminated groundwater in Ukraine. *Central Eur. J. Occup. Environ. Med.* **2009**, *15*, 73–85.
3. Wireman, M.; Kornilovych, B. Installation of a permeable reactive barrier in uranium mining district—East Central Ukraine. *Newsl. Int. Assoc. Hydrogeol. U.S. Natl. Chapter* **2012**, *41*, 12–14.
4. Interstate Technology & Regulatory Council. *Permeable Reactive Barriers: Lessons Learned/New Directions. PRB-4*; Interstate Technology & Regulatory Council, Permeable Reactive Barriers Team: Washington, DC, USA, 2005. Available online: www.itrcweb.org (accessed on 31 March 2018).
5. Interstate Technology & Regulatory Council. *Permeable Reactive Barriers: Technology Update. PRB-5*; Interstate Technology & Regulatory Council, PRB; Technology Update Team: Washington, DC, USA, 2011. Available online: www.itrcweb.org (accessed on 31 March 2018).
6. Fernando, W.A.M.; Ilankoon, I.M.S.K.; Syed, T.H.; Yellishetty, M. Challenges and opportunities in the removal of sulphate ions in contaminated mine water: A review. *Miner. Eng.* **2018**, *117*, 74–90. [CrossRef]
7. Fu, F.; Dionysiou, D.D.; Liu, H. The use of zero-valent iron for groundwater remediation and wastewater treatment: A review. *J. Hazard. Mater.* **2014**, *267*, 194–205. [CrossRef] [PubMed]

8. Waybrant, K.R.; Blowes, D.W.; Ptacek, C.J. Selection of reactive mixtures for use in permeable reactive walls for treatment of mine drainage. *Environ. Sci. Technol.* **1998**, *32*, 1972–1979. [CrossRef]
9. Gu, B.; Liang, L.; Dickey, M.J.; Yin, X.; Dai, S. Reductive precipitation of uranium (VI) by zero-valent iron. *Environ. Sci. Technol.* **1998**, *32*, 3366–3373. [CrossRef]
10. Bowman, R.S. Applications of surfactant-modified zeolites to environmental remediation. *Micropor. Mesopor. Mater.* **2003**, *61*, 43–56. [CrossRef]
11. Natale, F.D.; Natale, M.D.; Greco, R.; Lancia, A.; Laudante, C.; Musmarra, D. Groundwater protection from cadmium contamination by permeable reactive barriers. *J. Hazard. Mater.* **2008**, *160*, 428–434. [CrossRef] [PubMed]
12. Blowes, D.W.; Ptacek, C.J.; Jambor, J.L. In situ remediation of chromate contaminated groundwater using permiable reactive walls: Laboratory studies. *Environ. Sci. Technol.* **1997**, *31*, 3348–3357. [CrossRef]
13. Farrell, J.; Bostick, W.D.; Jarabek, R.J.; Fiedor, J.N. Uranium removal from ground water using zero-valent iron media. *Ground Water* **1999**, *37*, 618–624. [CrossRef]
14. Cundy, A.B.; Hopkinson, L.; Whitby, R.L.D. Use of iron-based technologies in contaminated land and groundwater remediation: A review. *Sci. Total Environ.* **2008**, *400*, 42–51. [CrossRef] [PubMed]
15. Wilkin, R.T.; Acree, S.D.; Ross, R.R.; Beak, D.G.; Lee, T.R. Performance of a zerovalent iron reactive barrier for the treatment of arsenic in groundwater: Part 1. Hydrogeochemical studies. *J. Contam. Hydrol.* **2009**, *106*, 1–14. [CrossRef] [PubMed]
16. Guo, X.; Yang, Z.; Don, H.; Guan, X.; Ren, Q.; Lv, X.; Jin, X. Simple combination of oxidants with zero-valent-iron (ZVI) achieved very rapid and highly efficient removal of heavy metals from water. *Water Res.* **2016**, *88*, 671–680. [CrossRef] [PubMed]
17. Li, S.; Wang, W.; Liang, F.; Zhang, W. Heavy metal removal using nanoscale zero-valent iron (nZVI): Theory and application. *J. Hazard. Mater.* **2017**, *322*, 163–171. [CrossRef] [PubMed]
18. Yabusaki, S.; Cantrell, K.; Sass, B.; Steefel, C. Multicomponent reactive transport in an in situ zero-valent iron cell. *Environ. Sci. Technol.* **2001**, *35*, 1493–1503. [CrossRef] [PubMed]
19. Jin, S.O.; Jeen, S.-W.; Gillham, R.W.; Gui, L. Effect of initial iron corrosion rate on long-term performance of iron permeable reactive barriers: Column experiments and numerical simulation. *J. Contam. Hydrol.* **2009**, *103*, 145–156. [CrossRef]
20. Chekli, L.; Bayatsarmadi, B.; Sekine, R.; Sarkar, B.; Maoz Shen, A.; Scheckel, K.G.; Skinner, W.; Naidu, R.; Shon, H.K.; Lombi, E.; et al. Analytical characterisation of nanoscale zero-valent iron: A methodological review. *Anal. Chim. Acta* **2016**, *903*, 13–35. [CrossRef] [PubMed]
21. Morrison, S.J.; Mushovic, P.S.; Niesen, P.L. Early breakthrough of molybdenum and uranium in a permeable reactive barrier. *Environ. Sci. Technol.* **2006**, *40*, 2018–2024. [CrossRef] [PubMed]
22. Lloyd, J.R.; Lovley, D.R. Microbial detoxication of metals and radionuclides. *Curr. Opin. Biotechnol.* **2001**, *12*, 248–253. [CrossRef]
23. Landa, E.R. Uranium mill tailings: Nuclear waste and natural laboratory for geochemical and radioecological investigations. *J. Environ. Radioact.* **2004**, *77*, 1–27. [CrossRef] [PubMed]
24. Malik, A. Metal bioremediation through growing cells. *Environ. Int.* **2004**, *30*, 261–278. [CrossRef] [PubMed]
25. Kalin, M.; Wheeler, W.N.; Meinrath, G. The removal of uranium from mining waste water using algal/microbial biomass. *J. Environ. Radioact.* **2005**, *78*, 151–177. [CrossRef] [PubMed]
26. Zagury, G.J.; Kulnieks, V.I.; Neculita, C.M. Characterization and reactivity assessment of organic substrates for sulphate-reducing bacteria in acid mine drainage treatment. *Chemosphere* **2006**, *64*, 944–954. [CrossRef] [PubMed]
27. Merroun, M.; Selenska-Pobell, S. Bacterial interactions with uranium: An environmental perspective. *J. Contam. Hydrol.* **2008**, *102*, 285–295. [CrossRef] [PubMed]
28. Benner, S.G.; Blowes, D.W.; Gould, W.D.; Gerbert, R.B.; Ptacek, C.J. Geochemistry of a permeable reactive barrier for metals and acid mine drainage. *Environ. Sci. Technol.* **1999**, *33*, 2793–2799. [CrossRef]
29. Logan, M.V.; Reardon, K.F.; Figueroa, L.A.; McLain, J.E.T.; Ahmann, D.M. Microbial community activities during establishment, performance, and decline of bench-scale passive treatment systems for mine drainage. *Water Res.* **2005**, *39*, 4537–4551. [CrossRef] [PubMed]
30. Qiu, R.; Zhao, B.; Liu, J.; Huang, X.; Li, Q.; Brewer, E.; Wang, S.; Shi, N. Sulfate reduction and copper precipitation by a *Citrobacter* sp. isolated from a mining area. *J. Hazard. Mater.* **2009**, *164*, 1310–1315. [CrossRef] [PubMed]

31. Spasonova, L.N.; Tobilko, V.Y.; Leshchuk, I.A.; Gvozdyak, P.I.; Kornilovich, B.Y. Sorption of U(VI) ions on the natural association of microorganisms of activated sludge and pure culture Bacillus polymyxa IMB 8910. *J. Water Chem. Technol.* **2006**, *28*, 61–68.
32. Chabalala, S.; Chirwa, E.M.N. Removal of uranium (VI) under aerobic and anaerobic conditions using an indigenous mine consortium. *Miner. Eng.* **2010**, *23*, 526–531. [CrossRef]
33. Zhengji, Y. Microbial removal of uranyl by sulfate reducing bacteria in the presence of Fe(III) (hydr) oxides. *J. Environ. Radioact.* **2010**, *101*, 700–705. [CrossRef] [PubMed]
34. Abdelouas, A.; Nuttall, H.E.; Lutze, W.; Lu, Y. In situ removal of uranium from groundwater. In Proceedings of the International Conference on Tailings and Mine Waste'98, Fort Collins, CO, USA, 26–28 January 1998; Balkema, A.A., Ed.; Brookfield, VT: Rotterdam, The Netherlands, 1998; pp. 669–677.
35. Abdelouas, A.; Lutze, W.; Gong, W.; Nuttall, H.E.; Strietelmeier, B.A.; Travis, B.J. Biological reduction of uranium in groundwater and subsurface soil. *Sci. Total Environ.* **2000**, *250*, 21–35. [CrossRef]
36. Spear, J.R.; Figueroa, L.A.; Honeyman, B.D. Modeling reduction of uranium U(VI) under variable sulfate concentrations by sulfate-reducing bacteria. *Appl. Environ. Microbiol.* **2000**, *66*, 3711–3721. [CrossRef] [PubMed]
37. Macaskie, L.E.; Bonthrone, K.M.; Yong, P.; Goddard, D.T. Enzymically mediated bioprecipitation of uranium by a *Citobacter* sp.: A concerted role for exocellular lipopolysaccharide and associated phosphatase in biomineral formation. *Microbiology* **2000**, *146*, 1855–1867. [CrossRef] [PubMed]
38. Bender, J.; Duff, M.C.; Phillips, P.; Hill, M. Bioremediation and bioreduction of dissolved U(VI) by microbial mat consortium supported on silica gel particles. *Environ. Sci. Technol.* **2000**, *34*, 3235–3241. [CrossRef]
39. Sasaki, K.; Blowes, D.W.; Ptacek, C.J.; Gould, W.D. Immobilization of Se(VI) in mine drainage by permeable reactive barriers: Column performance. *Appl. Geochem.* **2008**, *23*, 1012–1022. [CrossRef]
40. Guo, Q.; Blowes, D.W. Biogeochemistry of two types of permeable reactive barriers, organic carbon and iron-bearing organic carbon for mine drainage treatment: Column experiments. *J. Contam. Hydrol.* **2009**, *107*, 128–139. [CrossRef] [PubMed]
41. ISO 5667-3:2003. *Water Quality-Sampling*; International Organization for Standardization: Geneva, Switzerland, 2003.
42. Surface Water Sampling. Available online: https://www.epa.gov/quality/surface-water-sampling (accessed on 20 May 2018).
43. *Standard Methods for the Examination of Water and Wastewater*, 22nd ed.; Bridgewater, L., Rice, E.W., Baird, R.B., Eaton, A.D., Clesceri, L.S., Eds.; APHA-AWWA-WEF: Washington, DC, USA, 2012; 1496p, ISBN 9780875530130 0875530133.
44. OGWDW. Available online: http://www.epa.gov/OGWDW (accessed on 13 May 2018).
45. Pratopo, M.I.; Moriyama, H.; Higashi, K. Carbonate complexation of neptunium (IV) and analogous complexation of ground-water uranium. *Radiochim. Acta* **1990**, *51*, 27–31. [CrossRef]
46. Schmeide, K.; Sachs, S.; Bubner, M.; Reich, T.; Heise, K.H.; Bernhard, G. Interaction of uranium(VI) with various modified and unmodified natural and synthetic humic substances studied by EXAFS and FTIR spectroscopy. *Inorg. Chim. Acta* **2003**, *351*, 133–140. [CrossRef]
47. Kornilovich, B.Y.; Pshinko, G.N.; Bogolepov, A.A. Effects of EDTA and NTA on sorption of U(VI) on the clay fraction of soil. *Radiochemistry* **2006**, *48*, 584–588. [CrossRef]
48. Pagnanelli, F.; Viggi, C.C.; Mainelli, S.; Toro, L. Assesment of solid reactive mixtures for the development of biological permeable reactive barriers. *J. Hazard. Mater.* **2009**, *170*, 998–1005. [CrossRef] [PubMed]
49. Postgate, J.R. *The Sulfate Reducing Bacteria*, 2nd ed.; Cambridge University Press: Cambridge, NY, USA, 1984; 208p, ISBN 0521257913.
50. Liu, A.; Liu, J.; Han, J.; Zhang, W. Evolution of nanoscale zero-valent iron (nZVI) in water: Microscopic and spectroscopic evidence on the formation of nano- and micro-structured iron oxides. *J. Hazard. Mater.* **2017**, *322*, 129–135. [CrossRef] [PubMed]
51. Lee, T.R.; Wilkin, R.T. Iron hydroxy carbonate formation in zerovalent iron permeable reactive barriers: Characterization and evaluation of phase stability. *J. Contam. Hydrol.* **2010**, *116*, 47–57. [CrossRef] [PubMed]
52. Langmuir, D. *Aqueous Environmental Geochemistry*; Prentice Hall: Upper Saddle River, NJ, USA, 1997; 600p, ISBN 0023674121.
53. Kornilovich, B.Y.; Pshinko, G.N.; Koval'chuk, I.A. Effect of Fulvic Acids on Sorption of U(VI) on Clay Minerals of Soils. *Radiochemistry* **2001**, *43*, 528–531. [CrossRef]

54. Kamolpornwijit, W.; Liang, L.; West, O.R.; Moline, G.R.; Sullivan, A.B. Preferential flow path development and its influence on long-term PRB performance: Column study. *J. Contam. Hydrol.* **2003**, *66*, 161–178. [CrossRef]
55. Ilankoon, I.M.S.K.; Neethling, S.J. Liquid spread mechanisms in packed beds and heaps. The separation of length and time scales due to particle porosity. *Miner. Eng.* **2016**, *86*, 130–139. [CrossRef]

Article

Neural Network Modeling for the Extraction of Rare Earth Elements from Eudialyte Concentrate by Dry Digestion and Leaching

Yiqian Ma [1], Srecko Stopic [1,*], Lars Gronen [2], Milovan Milivojevic [3], Srdjan Obradovic [3] and Bernd Friedrich [1]

1 IME Process Metallurgy and Metal Recycling, RWTH Aachen University, Intzestraße 3, 52056 Aachen, Germany; yma@ime-aachen.de (Y.M.); bfriedrich@ime-aachen.de (B.F.)
2 Institute of Applied Mineralogy and Economic Geology (IML), RWTH Aachen University, Wüllnerstraße 2, 52062 Aachen, Germany; gronen@emr.rwth-aachen.de
3 Department of Informatics, Business and Technical College of Applied Studies, 31000 Uzice, Serbia; milovan.milivojevic@vpts.edu.rs (M.M.); srdjan.obradovic@sbb.rs (S.O.)
* Correspondence: sstopic@ime-aachen.de; Tel.: +49-2419-5860

Received: 8 March 2018; Accepted: 10 April 2018; Published: 13 April 2018

Abstract: Eudialyte is a promising mineral for rare earth elements (REE) extraction due to its good solubility in acid, low radioactive, and relatively high content of REE. In this paper, a two stage hydrometallurgical treatment of eudialyte concentrate was studied: dry digestion with hydrochloric acid and leaching with water. The hydrochloric acid for dry digestion to eudialyte concentrate ratio, mass of water for leaching to mass of eudialyte concentrate ratio, leaching temperature and leaching time as the predictor variables, and the total rare earth elements (TREE) extraction efficiency as the response were considered. After experimental work in laboratory conditions, according to design of experiment theory (DoE), the modeling process was performed using Multiple Linear Regression (MLR), Stepwise Regression (SWR), and Artificial Neural Network (ANN). The ANN model of REE extraction was adopted. Additional tests showed that values predicted by the neural network model were in very good agreement with the experimental results. Finally, the experiments were performed on a scaled up system under optimal conditions that were predicted by the adopted ANN model. Results at the scale-up plant confirmed the results that were obtained in the laboratory.

Keywords: eudialyte; rare earth elements; dry digestion; leaching; neural network

1. Introduction

Rare earth elements (REE) consist of scandium, yttrium, and the group of 15 lanthanide elements, which are characterized by similar chemical properties and commonly occur in the same deposits [1]. Currently, REE are vital component of many modern technologies, including electric and conventional cars, computers and smart phones, renewable energy infrastructure, and phosphor light [2–4]. Their unique properties, such as radiation emission or magnetism, allow for REEs to be used in many different therapeutic and diagnostic applications in modern medicine.

Due to the enormous significance that REEs will have in the future, research that is related to REEs is essential in many aspects: mineralogy, location of the world's REE deposits, classic and alternative extraction technologies, etc. Concerning mineralogy, the conventional primary minerals involve bastnaesite, monazite, xenotime, REE phosphate minerals, and ion adsorption minerals. The authors in [5] are focused on REE Mineralogy and Resources, which might be used for the recovery of Heavy rare earth elements concerning to the future green technologies. Bayan Obo in China and Mountain Pass in USA possess the largest carbonatite deposits around the world, which show light REE enrichment.

Ion-adsorption REE ore deposits that are mainly distributed in southern China, and the grade of heavy REE is high in it. Other ore deposits tend to be distributed unevenly globally [6,7]. Nowadays, above 90% of the REE in the world market is supplied by China, which has abundant rare earth mines. Demand for REE has been constantly increasing in last decades and is expected to grow even stronger in the future [8,9]. As such, many projects have been initiated all over the world, which seek to find alternative sources of REE outside China [10–12]. Except the secondary resources, a variety of potential unconventional deposits have also been explored, such as REE-bearing laterite, peralkaline granite, ferromanganese nodules, and silicate deposits, including eudialyte, mosandrite, and britholite [13–16]. Among the minerals mentioned, Eudialyte is a promising REE resource, which contains 1.5–2% REE [17]. Eudialyte is a somewhat complicated cyclosilicate mineral, which forms in alkaline igneous rocks, such as nepheline syenites. It tends to be rich in zirconium, beryllium, cerium, niobium, barium, yttrium, and other rare earth elements [18,19]. The typical chemical formula of Eudialyte is $Na_4(Ca, Ce, Fe)_2ZrSi_6O_{17}(OH, Cl)_2$, but the component of the mineral may be sometimes different because of the zeolite crystal structure, which possesses ion-exchange properties of eudialyte [20,21].

The technologies for the REE recovery vary depending upon the characteristics of initial material. Using acid, alkali, or chemical salts to decompose and extract represent the common ones. For example, REE in ion-adsorption ore can be readily extracted with ammonia sulfate solution. To extract REE from eudialyte, technological strategies, like decomposition with sulfuric, hydrochloric, and nitric acids have already been suggested and partly researched in the laboratory. Despite being easily decomposed with acids, the treatment can be very challenging as co-dissolved silica forms a gelatinous mass of the whole suspension [22–24]. Russian teams [22,23] studied the hydrometallurgical treatment of eudialyte in an earlier period, and the advisable technology would be a two stage decomposition process by sulfuric acid. Sodium sulfate solution was used to dilute the slurry after the first decomposition by acid, keeping REE staying in the solid as double sulfate salts. REE were recovered in the subsequent stage, washing and converting the sulfates into nitrates or chlorides. However, it required much excess acid to avoid the formation of silica gel [22,25]. Another choice is reaction with hydrogen fluoride [23]. It was pointed out fluoride ion can facilitate the decomposition of eudialyte as well as prevent gel formation. The major drawback of this process is safety concern as HF is used. After using a beneficiation process (a combination of gravity and magnetic separation), the authors of [26] studied hydrometallurgical processing of eudialyte Norra Kärr (Sweden) through various leaching experiments with sulfuric acid. In this project, dry digestion with HCl was not applied. Rivera et al. have studied the extraction of rare earth elements from bauxite residue (red mud) by dry digestion followed by water leaching through multi-stage circulation of acid in small laboratory scale [27], but mathematical modeling and process optimization was not performed.

According to the latest research [17], fuming ("Dry digestion at temperatures about 100 °C"), the addition of an acidic solution to heated eudialyte concentrate, and the subsequent water leaching of the treated concentrate could result to the high recovery of REE with avoiding silica gel formation. Moreover, in the mentioned paper, Davris et al. studied hydrometallurgical processing of eudialyte bearing concentrates, using high concentrated acid to recover rare earth elements via low-temperature dry digestion to prevent silica gel formation. Both fuming and low-temperature dry digestion processes confirmed HCl performs higher recovery of REE when compared with H_2SO_4. It was found that dry digestion with high concentrated HCl benefits the extraction of REE and prevents gel formation. The recommended concentration of HCl to decompose the eudialyte concentrate is 10 mol/L [24].

In this paper, the dry digestion process was studied without external heating. In order to reduce the number of complex laboratory experiments, the design of experiments theory (DOE) was firstly used in order to create an optimal central composite design plan (CCD). This approach for the extraction of rare earth elements is missing in literature, and is not present in the previous work of Voßenkaul [24]. In addition, information on upscaling operations for the treatment of eudialyte is missing in overall literature. In order to obtain the optimal regime for the REE extraction process, based on the realized DOE plan, we performed modeling of this process via Multiple Linear Regression (MLR), Stepwise

Regression (SWR), and Artificial Neural Network (ANN). To our knowledge, no previous work has been reported on ANN modeling for the extraction of rare earth elements from eudialyte concentrate by dry digestion and leaching. Our goal was to verify the feasibility of low temperature acid treatment of eudialyte concentrate on a scale-up demonstration plant using the obtained optimal set of process parameters.

2. Theoretical Background

Theoretical background of applied methods—MLR, SWR, DOE, and ANN—is shown in the following subsections.

2.1. Multiple Linear Regression

MLR is used to model the linear relationship between a response variable and one or more predictor variables. MLR model represents a linear combination of regression coefficients β_i and regressors, in the form of $\mathbf{X} = f_i(\mathbf{x}^\mathbf{T})$, named basis functions or states: $\eta = \eta(\boldsymbol{\beta}, \mathbf{x}) = \sum_{i=0}^{d} \beta_i \cdot f_i(\mathbf{x}^\mathbf{T})$. Vector $\mathbf{x}$ is a vector of predictors, $\boldsymbol{\beta}$ is a vector of regression coefficients, and d is the number of basis functions. In MLR, the unknown coefficients $\beta_i (i = 0, d)$ are estimated by values $b_i (i = 0, d)$, which were obtained by using experimental results. In matrix notation, the least squares estimator of $\boldsymbol{\beta}$ is $b = (\mathbf{X}^\mathbf{T} \cdot \mathbf{X})^{-1} \cdot \mathbf{X}^\mathbf{T} \cdot \mathbf{Y}$. Vector $\mathbf{Y}$ is a vector of output variables, the fitted model is given by $\hat{\mathbf{Y}} = \mathbf{X} \cdot \mathbf{b}$, and the vector of the residuals is $\boldsymbol{\varepsilon} = \mathbf{Y} - \hat{\mathbf{Y}}$ [28–30].

2.2. Stepwise Regression

Stepwise regression is essentially based on semi-partial correlation, which is expressed through the semi-partial correlation coefficient (sr) and the square of this coefficient (sr^2). The square of the semi-partial correlation coefficient, for a specific single variable, indicates by how much the R^2 value will reduce if this single variable is removed from the regression equation. R^2 denotes the coefficient of multiple determination.

Let χ be the set of all the independent variables $\mathbf{X}$, and ψ_k be the set of all the independent variables $\mathbf{X}$ except for x_k. Then, the squared semi-partial correlation coefficient is expressed as follows in (1):

$$sr_k^2 = R_\chi^2 - R_{\psi_k}^2 \tag{1}$$

Form of expressing semi-partial correlation coefficients may vary, and one of the most common forms is (2):

$$sr_k = \frac{t_k \cdot \sqrt{1 - R_\chi^2}}{\sqrt{residualDF}} \tag{2}$$

where t_k is the Student's t-statistic value for the kth regressor in the MLR model, $residualDF = N - K - 1$, denotes the number of degrees of freedom for the sum of residuals, N is number of measurements, and K is the number of regressors. Stepwise regression is explained in detail in [28,30,31].

2.3. Design of Experiments

Design of experiment (DOE) was performed in order to decrease the number of necessary experiments, required time, and operational costs. This method stems from the studies of Fisher [32] and Box and Wilson [33]. With DOE, the research hyperspace can be explored with high efficiency, factor interactions can be revealed, and a substantial volume of response space can be covered. In this paradigm, D-optimality criterion is the basis for the design matrix, as denoted by $\mathbf{X}$, which contains a set of experimental points, which lie on the border of the hyper sphere, forming a hypercube [28].

In order to make this analysis simple, the process variables associated with factors X_i are encoded using the following equations,

$$x_i = \frac{X_i - X_{0i}}{w_i}, (i = 1, 2, 3 \ldots, k) \tag{3}$$

$$X_{0i} = \frac{X_{i\max} + X_{i\min}}{2}, w_i = \frac{X_{\max} - X_{\min}}{2 \cdot \alpha} \tag{4}$$

where k denotes the number of significant factors, X_i is an actual value of the ith factor, X_0 represents the arithmetic mean for factor X_i (basic level), and w_i is the interval of its variation. The actual value of *alpha* depends of the number of factors: $alpha = 2^{k/4}$ and has a value of 1.41 for two factors, 1.682 for three factors, etc. For example, an experimental plan with three factors ($k = 3$) for first order MLR, in schematic form is given in Figure 1a. In the encoded domain, the highest factor level corresponds to +1, the lowest level to -1, and the basic level to 0. The grey point in the center of the cube is the center point of the experiment design. It is usually expected that the performed experiments shall be repeated in the central point in order to determine the experimental error [28].

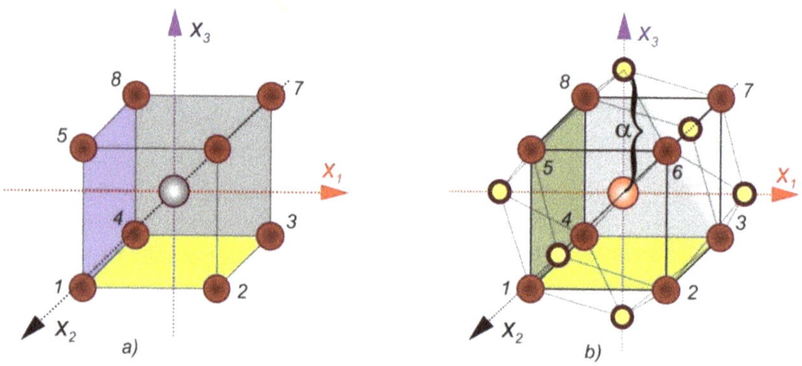

Figure 1. Arrangement of experimental points in 2^k space: (**a**) k = 3; (**b**) k = 3, *alpha* = 1.682.

Polynomial regression models in the second-order MLR form require additional experimental points. For full factorial orthogonal rotatable central composite design (CCD), additional experimental axial points are at the distance alpha from the central point. Schematic view of design of experiments for this kind of model is shown in Figure 1b.

CCD factorial design is a common approach for research optimization with a limited number of experiments [34]. MLR models of higher order require additional measurements. However, additional experiments would increase the investigation costs.

2.4. Artificial Neural Network

Among many different types of ANN, the feed-forward ANN is the most widely researched and used [35–37]. A feed-forward network has many layers. The structure of a feed-forward ANN is given in Figure 2.

In Figure 2 $\mathbf{X}$ denotes the vector of input variables, $\mathbf{Y}$ is the vector of output variables, $\mathbf{w}^{(\mathbf{i},\mathbf{h})}$ is a column-matrix of weight coefficients between neurons in the input layer and the first hidden layer, and $\mathbf{w}^{(\mathbf{h},\mathbf{o})}$ is a column-matrix of weight coefficients between the neurons in the last hidden layer and the output layer. The term $w_{ij}^{(q-1,q)}$ denotes the weight between ith neuron from $(q-1)$th hidden layer and jth neuron from (q)th hidden layer.

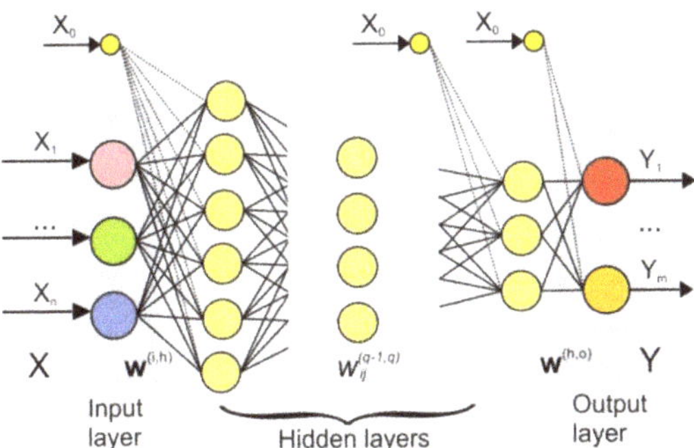

Figure 2. Schematic view of a feed-forward Artificial Neural Network (ANN).

The total input $\alpha_j^{(q)}$ into processing element j in qth layer is a weighted sum of all the outputs $x_i^{(q-1)}$ from the previous layer. The activation function, $x_j^{(q)} = AF(\alpha_j^{(q)})$, transforms the input signal, $\alpha_j^{(q)}$, to the output signal—$x_j^{(q)}$. Most frequently used activation functions are: logistic sigmoid, hyperbolic tangent, and rectified linear unit.

Values of all the ANN weight coefficients are determined using a learning process. The goal is to minimize the error function (difference between desired and predicted values). One of the most popular ANN learning algorithms is the backpropagation algorithm. According to this algorithm, the error is propagated backward through the network and the weights are updated through a number of iterations (Figure 3). The procedure progresses until it reaches convergence or a given maximum number of iterations [38].

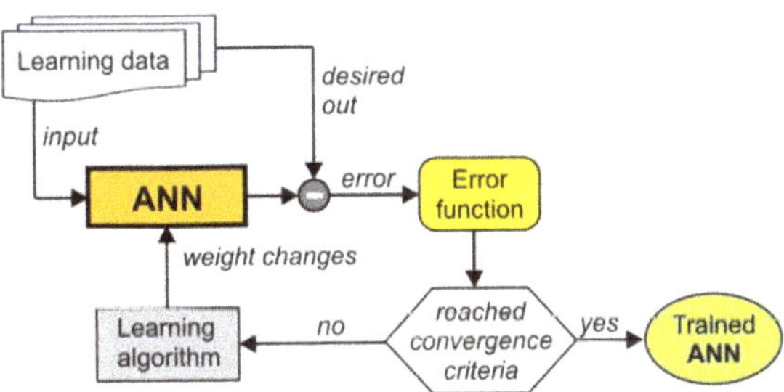

Figure 3. Supervised learning model.

In this work we used some commonly used variations of the backpropagation algorithm: The Backpropagation gradient descent algorithm (BPGD) [39,40], the Resilient backpropagation algorithm (RPROP) [41,42], and adaptive moment estimation algorithm (ADAM) [43].

Design of a neural network is determined by various ANN hyperparameters: number of hidden layers, number of neurons in each hidden layer, number of training samples, initial set of weights, the range of initial weights, etc. Some recommendations for hyperparameter choice are given in [44].

3. Materials and Methods

3.1. Material and Analysis

Eudialyte concentrate from Tanbreez ore exploited in southern Greenland was used as raw material. The name Tanbreez is the acronym of Ta, Nb, REE, and Zr. The chemical composition of the eudialyte concentrate supplied by Geological Survey of Finland (GTK) is given in Table 1. The composition of eudialyte concentrate and solid residue were analyzed by X-ray fluorescence (AXIOS, PANanalytical, Eindhoven, The Netherland) and the elements in the solution were analyzed by ICP-OES (SPECTRO ARCOS, SPECTRO Analytical Instruments GmbH, Kleve, Germany). The QEMSCAN results in Figure 4 reveal the mineral distribution of eudialyte concentrate by false red color image and the right is the BSE image. The mineral composition and the microstructure were determined by QEMSCAN (Quantitative Evaluation of Materials by SCANning electron microscopy), (FEI/Thermo corporate headquarters, North America Nanoport, Hillsboro, OR, USA). The phase quantification is summarized in Figure 5. Eudialyte accounts for 67.05% in the eudialyte concentrate and the concentrate contains some other silicate minerals, such as Arfvedsonite, Nepheline, and Feldspar.

Commercial grade HCl (31%) with molarity of 10 mol/L and tap water without further purification were used, both in the laboratory and scale-up plant.

Table 1. Chemical composition of eudialyte concentrate by XRF (X-ray fluorescence).

Element	Content (wt %)	Element	Content (wt %)
Al	3.2	Ce	0.52
Ca	5.7	Pr	532 mg/kg
Fe	6.04	Nd	0.21
Mn	0.39	Sm	440 mg/kg
Nb	0.36	Gd	239 mg/kg
Zr	5.08	Dy	580 mg/kg
Hf	0.11	Y	0.33
Si	23.1	Yb	368 mg/kg
La	0.25	TREE	1.52

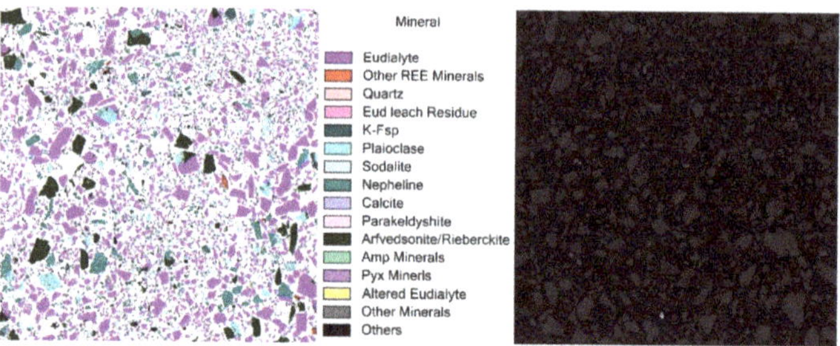

Figure 4. QEMSCAN analysis of eudialyte concentrate ((**left**) false color image after evaluation, (**right**) initial BSE-image).

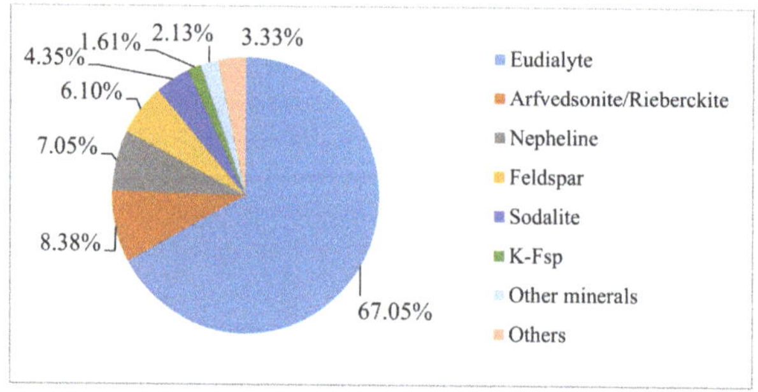

Figure 5. Mineral distribution of eudialyte concentrate (analysis by GTK, Finnland).

3.2. Extraction Procedure

The operation flowchart is shown in Figure 6. Dry digestion process with high concentrated 31% hydrochloric acid represented the first step. The molarity of the used acid was 10 mol/L. Taking advantages of the exothermic reaction, the temperature could reach about 80 °C without external heating. The aim was to decompose eudialyte concentrate and form metal chloride salts without silica gel formation by forming silica precipitation. The specified amount of water was subsequently added to the sludge after digestion, achieving REE leaching and allowing low silica to be dissolved into the leaching solution. Therefore, the maximum REE extraction process time (dry digestion and leaching) in the laboratory was 2 h. After filtration and washing the filter cake, a REE enriched solution and solid residue were obtained.

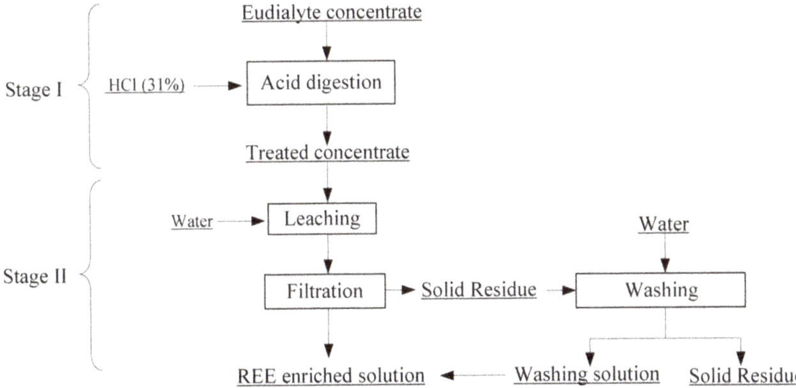

Figure 6. Proposed flowchart of the treatment of rare earth elements (REE) extraction from eudialyte concentrate.

4. Experiments and Obtained Measurements of Process Outputs

During research we chose to investigate four factors ($k = 4$), namely hydrochloric acid to the concentrate ratio (cHCl), water to the concentrate ratio (cH_2O), leaching temperature (T), and leaching time (t). The chosen parameters were proposed concerning the previous experimental work at the EURARE Project [11], literature data [17,24], and high experience of the authors in hydrometallurgy.

The arrangement of experimental points for $k \geq 4$ cannot be visualized, but it is easily described using a design matrix. In accordance with CCD plan, the central point experiments were repeated seven times to guarantee a more precise determination of an experimental error. Eudialyte concentrate mass in all laboratory experiments was 500 g. The total experimental design matrix and the obtained results are given in Table 2.

Table 2. Experiment design matrix.

No.	X1:HCl:Con. (L:kg)		X2:Water:Con. (L:kg)		X3:Leaching Temp. (°C)		X4:Leaching Time (min.)		Y:TREE Extr. (%)
	1:1, 1.25:1, 1.5:1		1:1, 2:1, 3:1		20, 50, 80		20, 40, 60		[0–100]
	c.v.	a.v.	c.v.	a.v.	c.v.	a.v.	c.v.	a.v.	
A1	1	1.5:1	1	3:1	1	20	1	60	93.475
A2	−1	1:1	1	3:1	1	20	1	60	79.640
A3	1	1.5:1	−1	1:1	1	20	1	60	82.010
A4	−1	1:1	−1	1:1	1	20	1	60	73.940
A5	1	1.5:1	1	3:1	−1	20	1	60	86.465
A6	−1	1:1	1	3:1	−1	20	1	60	79.070
A7	1	1.5:1	−1	1:1	−1	20	1	60	81.170
A8	−1	1:1	−1	1:1	−1	20	1	60	79.450

Table 2. *Cont.*

No.	X1:HCl:Con. (L:kg)		X2:Water:Con. (L:kg)		X3:Leaching Temp. (°C)		X4:Leaching Time (min.)		Y:TREE Extr. (%)
	1:1, 1.25:1, 1.5:1		1:1, 2:1, 3:1		20, 50, 80		20, 40, 60		[0–100]
	c.v.	a.v.	c.v.	a.v.	c.v.	a.v.	c.v.	a.v.	
A9	1	1.5:1	1	3:1	1	20	−1	20	85.350
A10	−1	1:1	1	3:1	1	20	−1	20	80.630
A11	1	1.5:1	−1	1:1	1	20	−1	20	76.940
A12	−1	1:1	−1	1:1	1	20	−1	20	71.140
A13	1	1.5:1	1	3:1	−1	20	−1	20	77.650
A14	−1	1:1	1	3:1	−1	20	−1	20	75.525
A15	1	1.5:1	−1	1:1	−1	20	−1	20	88.600
A16	−1	1:1	−1	1:1	−1	20	−1	20	80.090
A17 *	0	1.25:1	0	2:1	0	50	1	40	89.49, 87.99, 88.74, 90.85, 86.63, 94.35, 83.24

Note: c.v., coded value; a.v., actual value; * repeated experiments in the central point.

The average total rare earth elements (TREE) extraction efficiency of all the experiments was over 80%, and the mean value of the central point A17 was 88.756% with a standard deviation of 3.458%.

Correlation analysis offers the following values of the Pearson between predictor variables: cHCl, $c\mathrm{H_2O}$, T and t at one side, and a dependent variable $\mathrm{TREE_{eff.}}$—TREE extraction efficiency, on the other side: $r_{\mathrm{TREE_{eff.}}-c\mathrm{HCl}} = 0.5631$, $r_{\mathrm{TREE_{eff.}}-c\mathrm{H_2O}} = 0.2640$, $r_{\mathrm{TREE_{eff.}}-T} = -0.0528$ and $r_{\mathrm{TREE_{eff.}}-t} = -0.2082$, respectively. According to advice in [45], the variables should be considered in a regression model only for $r > 0.3$. However, in order to perform a detailed analysis, the relationship of the leaching temperature and total rare earth elements extraction efficiency is also studied.

5. Model Setup and Results

5.1. Process Modeling in MLR Form

In this paper, we used the following MLR models:

(a) first order multiple linear regression

$$y = b_0 + \sum_{i=1}^{k} b_i \cdot x_i \tag{5}$$

(b) first order multiple linear regression with interaction effects

$$y = b_0 + \sum_{i=1}^{k} b_i \cdot x_i + \sum_{i=1}^{k-1} \sum_{j=i+1}^{k} b_{ij} \cdot x_i \cdot x_j + \sum_{i=1}^{k-2} \sum_{j=k+1}^{k-1} \sum_{l=j+1}^{k} b_{ijl} \cdot x_i \cdot x_j \cdot x_k \tag{6}$$

where k denotes the number of input factors ($k = 4$).

The statistical data analysis in this paper was performed in the R programming language [46]. Model adequacy was tested using Fisher's F-test. Model given by (5) was adequate, according to the F-test. Its regression parameters, together with its 95% confidence intervals are given in Table 3.

Table 3. Multiple Linear Regression (MLR) models, regression coefficients and their 95% confidence interval, F-test values, as well as the standard deviation of residuals σ_{res}.

First-Order MLR Model: $y=\beta_0+\beta_1 \cdot x_1+\beta_2 \cdot x_2+\beta_3 \cdot x_3+\beta_4 \cdot x_4$			
β_0	83.144565	[81.378 – 84.911]	$F_t(f_R, f_E, \alpha) = F_t(12, 6, 0.05) = 4$
β_1	3.2609375	[1.143 – 5.379]	$F_{rLF} = 3.90$
β_2	1.5290625	[−0.589 – 3.647]	$s^2(y) = 11.958262;\ s(y) = 3.458072$
β_3	−0.3059375	[−2.424 – 1.812]	$s^2{}_{LF} = 46.61679$
β_4	1.2059375	[−0.912 – 3.324]	$F_{rLF} < F_t$ *is adequate*, $\sigma_{res} = 6.8276488$

The analytical form of the adequate model in the space of encoded variables is the following:

$$y = 83.144565 + 3.2609375 \cdot x_1 + 1.5290625 \cdot x_2 - 0.3059375 \cdot x_3 + 1.2059375 \cdot x_4 \tag{7}$$

After transforming the predictor variables back into their original value domain, the model takes the form of (8) and its graphic representation, for (a) $T = const. = 20\ ^\circ\mathrm{C}$, $t = 20$ min. and (b) $T = const. = 20\ ^\circ\mathrm{C}$, $t = 60$ min. is given in Figure 7.

$$\mathrm{TREE_{eff.}} = 59.9058 + 13.0437 \cdot c\mathrm{HCl} + 1.5291 \cdot c\mathrm{H_2O} - 0.010197 \cdot T + 0.0602969 \cdot t \tag{8}$$

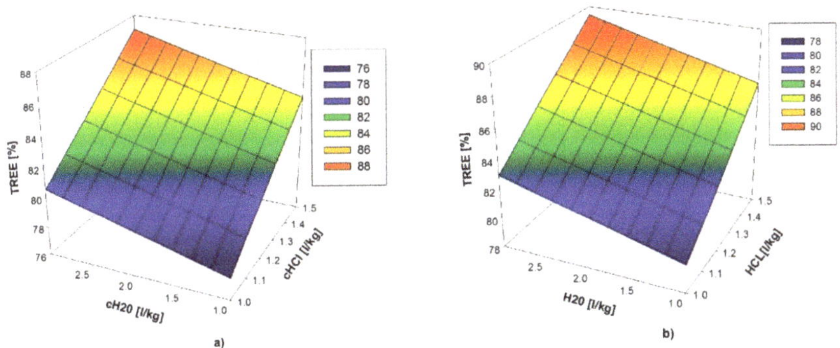

Figure 7. $\mathrm{TREE_{eff.}} = f(c\mathrm{HCl}, c\mathrm{H_2O})$ for (**a**) $T = const. = 20\ ^\circ\mathrm{C}$, $t = const. = 20$ min.; and, (**b**) $T = const. = 20\ ^\circ\mathrm{C}$, $t = const. = 60$ min.

Model (8) confirms the results of correlation analysis, that the leaching temperature (T) is not significant. Its changes throughout the range from 20 °C to 80 °C, cause a decrease of TREE extraction percentage only in a very small range from 0.20 to 0.81%, respectively. Figure 7 shows that the percentage of TREE extraction increases as the concentrations of HCl (variable cHCl) and H_2O (variable cH_2O) rise, and that the maximum values of extraction are around 90%. Furthermore, TREE extraction efficiency increases with the extension of time from 20 min. (Figure 7a), to 60 min. (Figure 7b). Although the model is in the form of a hyperplane (7), i.e., (8), adequate ($F_{rLF} = 3.9 < F_t = 4$), it can be observed that the process is characterized by high levels of noise, which is indicated by the standard error ($s(y) = 3.458072$). For this standard error value, it is relatively easy to perform proof of adequacy. In addition to the above, although the model (8) is adequate, its accuracy is very low, as can be seen from the value of its parameters: the coefficient of multiple determination ($R^2 = 0.433$), the adjusted coefficient of multiple determination ($R^2_{adj} = 0.244$), and root mean squared error, which is calculated in total for all of the experiments, is $RMSE = 4.668$.

This implies the need to describe the process with additional models of class (6). However, the potential number of such models is over 16,000, so the computational costs for modeling via the brute force method would be too high. For this purpose, the following section uses the SWR paradigm. Second order MLR models are not used, since that would require the expansion of the experimental plan with new experiments, according to CCD strategy.

5.2. Modeling with Stepwise Regression

For modeling, which is based on stepwise regression method, SPSS Statistics v24 software (IBM, Armonk, NY, USA) package was used. The initial set of regressors included variables: cHCl, cH_2O, T, t, $c\mathrm{HCl} \cdot c\mathrm{H_2O}$, $c\mathrm{HCl} \cdot T$, $c\mathrm{HCl} \cdot t$, $c\mathrm{H_2O} \cdot T$, $c\mathrm{H_2O} \cdot t$, $T \cdot t$, $c\mathrm{HCl} \cdot c\mathrm{H_2O} \cdot T$, $c\mathrm{HCl} \cdot c\mathrm{H_2O} \cdot t$, $c\mathrm{HCl} \cdot T \cdot t$, and $c\mathrm{H_2O} \cdot T \cdot t$. These regressors correspond to linear first order models in the form of Equation (6).

Model selection method, Forward selection was used, and the Akaike Information Criterion (AICC) was used as the optimality criterion for entry and removal of regressors into or from the model. In the iterative stepwise procedure, a model (2) from Table 4 with accuracy: $R^2_{adj} = 0.413$, $RMSE = 4.487$ was selected. When considering model (8), we deemed the alternative model (1) an oversimplification of the TREE extraction process. Model (2) satisfies the condition of non–multicollinearity (Tolerance > 0.1 and VIF < 10, Table 4).

Table 4. Linear regression models generated by means of stepwise procedure.

Model		Unstandard. Coeff.		Standard. Coeff.	t	Sig.	95.0% Confidence Interval for B		Correlations			Collinearity Statistics	
		B	Std. Error	Beta			Lower Bound	Upper Bound	Zero-Order	Partial	Part	Tolerance	VIF
1	(Constant)	64.866	6.293		10.307	0.000	51.452	78.280					
	cHCl	13.044	4.943	0.563	2.639	0.019	2.509	23.579	0.563	0.563	0.563	1.000	1.000
2	(Constant)	64.866	5.651		11.478	0.000	52.746	76.986					
	cHCl	10.554	4.587	0.456	2.301	0.037	0.715	20.393	0.563	0.524	0.441	0.936	1.068
	$c\mathrm{HCl}\cdot c\mathrm{H_2O}\cdot t$	0.031	0.015	0.425	2.146	0.050	0.000	0.062	0.540	0.497	0.411	0.936	1.068

[a] Dependent Variable: $\mathrm{TREE_{eff.}}$ [%].

In Table 4, the column Standardized Coefficients Beta shows the influence of the regressors on the dependent variable. Its values represent the coefficients of the regression model, normalized to have zero mean and unit variance. Absolute values of these coefficients describe the influence of individual regressors on the dependent variable $\mathrm{TREE_{eff}}$. From column *Sig.* it can be seen that both regressors (cHCl, $c\mathrm{HCl}\cdot c\mathrm{H_2O}\cdot t$) give a unique contribution to the model because the corresponding values are less than or equal to 0.05. The magnitudes of individual contributions are expressed through semipartial correlation coefficients, Part, i.e., the squared values of these coefficients ($0.441^2 = 0.194481$, $0.411^2 = 0.168921$). This further extends the understanding of the significance of the observed regressors, and the sum of Part usually differs from R^2_{adj} due to existence of joint effects.

The analytical form of the selected model is given by Equation (9) and its visual representation, for $t = const. = 60$ min., is given in Figure 8a:

$$\mathrm{TREE_{eff.}} = 64.86596 + 10.55357 \cdot c\mathrm{HCl} + 0.0312 \cdot c\mathrm{HCl} \cdot c\mathrm{H_2O} \cdot t \quad (9)$$

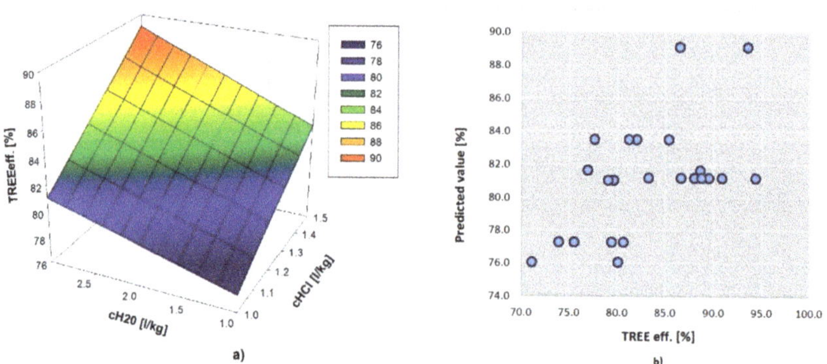

Figure 8. (**a**) Total rare earth elements (TREE) extraction model; and, (**b**) Predicted by Observed plot for $\mathrm{TREE_{eff}}$.

The final parameters of the model are: AICC = 55.213, and the predictor importances are 0.53 and 0.47 for cHCl and $c\mathrm{HCl}\cdot c\mathrm{H_2O}\cdot t$, respectively. Plot Predicted By Observed (Figure 8b), shows the predicted (fitted) values for the output variable $TREE_{eff.}$ [%], on the vertical axis and true values of the output variable on the horizontal axis. For a perfect model fit, points on the plot would lie on the diagonal line at an angle of 45°.

Like the linear model (8), the linear model with "interaction" variables (9) is adequate ($F_t(f_R, f_E, \alpha) = F_t(14, 6, 0.05) = 3.956$, $F_{rLF} = 3.17$, $F_{rLF} < F_t$ is adeuqate, $\sigma_{res} = 6.1580326$). Model (9) does not contain the *T* predictor, which confirms previous conclusions that the leaching temperature is not significant. Comparison of models (8) and (9) shows that both models are of similar shape (Figures 7b and 8a), but that the model with interactions (9) is a bit more accurate ($RMSE_{(9)} = 4.487$, $R^2_{adj(9)} = 0.413$, in comparison with $RMSE_{(8)} = 4.668$, $R^2_{adj(8)} = 0.244$).

Linear models (8) and (9) are adequate. However, the relatively low accuracy of the models, the complexity of the REE extraction process as well as the high level of dispersion, which can be observed in Figure 8b, suggest the possibility that the modeled extraction process is non-linear. In this context, the following section considers the application of the ANN paradigm for modeling the given process.

5.3. Modeling with ANN

5.3.1. ANN Modeling of TREE Extraction Based on LOO CV

Modeling by artificial neural networks requires a larger amount of data, or a larger number of measurements, than a set of experimental points that were generated by the CCD plan in this study. However, increasing the number of experiments could significantly increase research costs and time. For these reasons, the leave-one-out cross validation (LOO CV) concept, which attempts to generate a stable ANN model from the available data, was applied. This procedure is shown in Figure 9.

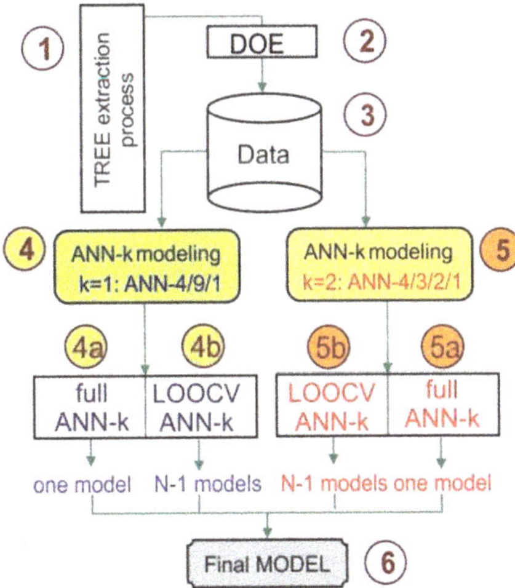

Figure 9. ANN TREE extraction modeling.

Data on the process of TREE extraction (1), based on the realized experiment plan (2), is stored in the database (3). In order to increase stability, modeling is carried out with two network topologies: (4) ANN-4/9/1 with four neurons in the input layer, one output, and nine neurons in the hidden layer, and (5) ANN-4/3/2/1 with four neurons in the input layer, one output, and two hidden layers with three and two neurons. Network architecture (4) is based on the recommendations given in [47], while the network architecture (5) is determined by the trial and error method. Schematic representations of these networks are given in Figure 10a,b.

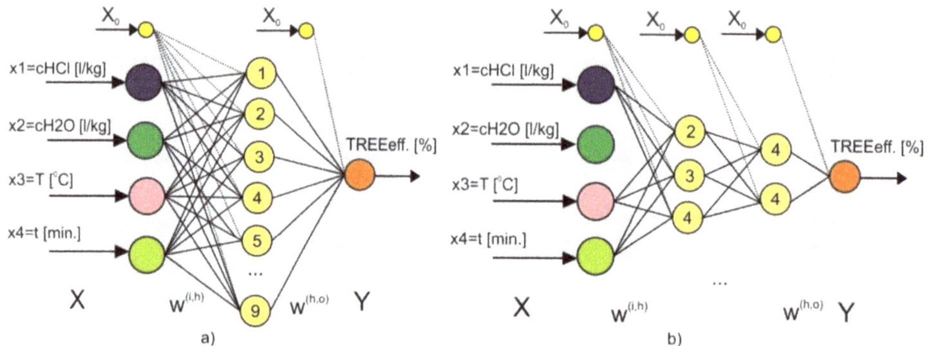

Figure 10. ANN topology for TREE modeling: (**a**) ANN-4/10/1; and, (**b**) ANN-4/3/2/1.

Each network architecture generates 17 ANN models ($N = 17$). Models *Full* $ANN_{k=1}$ (4a) and *Full* $ANN_{k=2}$ (5a) were created on the basis of all 17 input/output pairs, where the 17th pair corresponds to the average measurement at the central point of the experimental plan (A17, Table 2). These models represent ANN models, which selected network architectures can generate. In addition, both network architectures (4) and (5) generate 16 ANN models ($N - 1 = 16$), based on the LOO CV strategy. Leave-one-out Cross validation is a model validation method, which is described in detail in [48]. For very sparse datasets, as in our research, LOO CV can be used in order to train the ANN on the largest possible number of examples. For a dataset with N examples, N ANN models are trained. For each model, $N - 1$ examples are used for learning (training) and the remaining example for ANN testing (Figure 11).

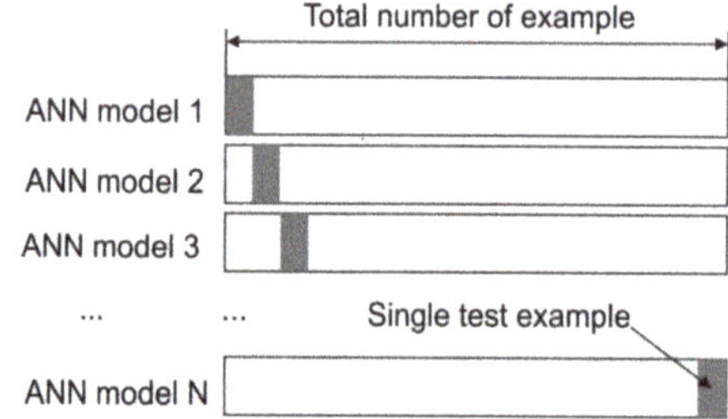

Figure 11. Leave-one-out Cross Validation in ANN modeling.

During each ANN modeling session, we arbitrarily retained in the training set the example that corresponds to the central point of the experiment (A17). In this way, units LOO CV $ANN_{k=1}$ (4b) and LOO CV $ANN_{k=2}$ (5b) each generated 16 ANN models, as mentioned earlier. Therefore, for ANN LOO CV modeling, only one sample is in the test dataset, but its exclusion from the training dataset significantly influences the ANN model's form. Due to space, as an example, Figure 12 shows only a few ANN models that were obtained during ANN LOO CV modeling.

The final model (6) on Figure 9, represents a data table numerical model, based on averaging 34 ANN models, which are generated in total by units (4) and (5).

Each $TREE_j$ point of the data table model (6) is calculated according to the Equation (10):

$$TREE_j = \frac{\sum_{k=1}^{2} TREE_{j,i}^{(Full,k)} + \sum_{i=1}^{N-1} \sum_{k=1}^{2} TREE_{j,i}^{(LOOCV,k)}}{2 \cdot N} \quad (10)$$

where $TREE_{j,i}^{(Full,k)}$ denoted *j*th point of *i*th ANN model produced by *Full* processes (4a) and (5a), while $TREE_{j,i}^{(LOOCV,k)}$ denoted *j*th point of *i*th ANN model produced by *LOOCV* iterative processes (4b) and (5b). In Equation (10), *k* denoted net topology: $k = 1$ and $k = 2$ for ANN-4/9/1 and ANN-4/3/2/1, respectively.

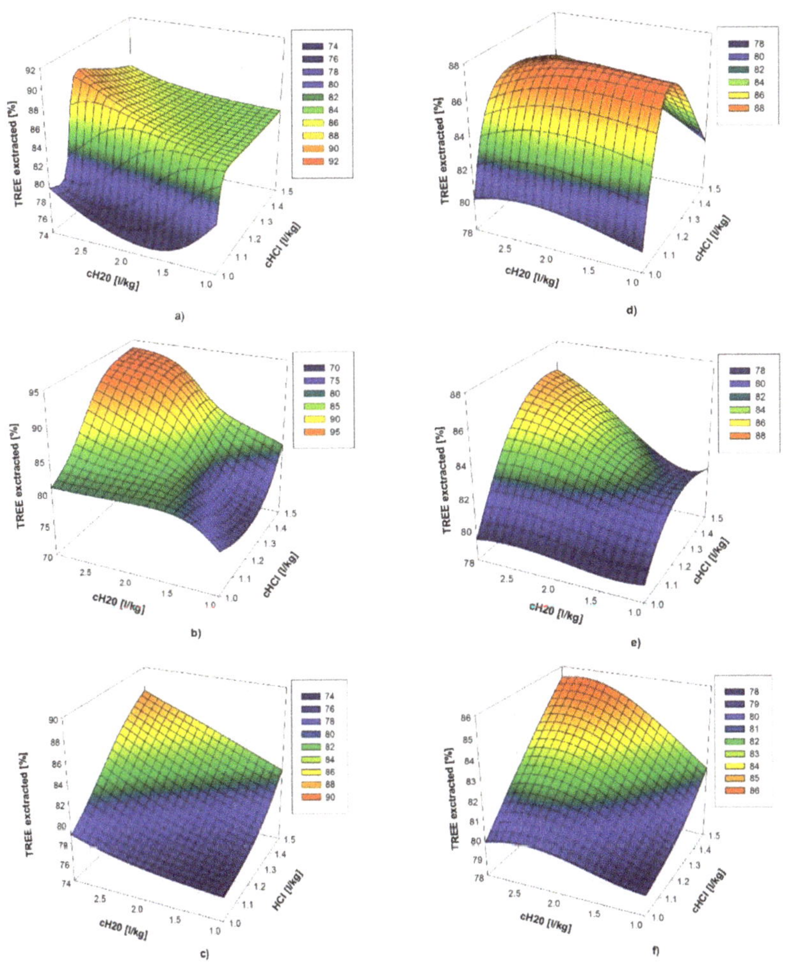

Figure 12. Examples of ANN models obtained during the leave-one-out cross validation (LOO CV) process: (**a**) ANN-4/3/2/1 model: excluded sample A4; (**b**) ANN-4/3/2/1 model: excluded sample A5; (**c**) ANN-4/3/2/1 model: excluded sample A10; (**d**) ANN-4/10/1 model: excluded sample A2; (**e**) ANN-4/10/1 model: excluded sample A10; and, (**f**) ANN-4/10/1 model: excluded sample A16.

5.3.2. Simulation and Prediction Based on ANN Models

ANN TREE extraction process modeling was implemented on R platform [46], using the neuralnet package. Leave-one Out Cross Validation strategy was implemented in R scripts that were written by authors. The developed ANN modeling strategy, as shown on Figure 10, is completely automated.

For unit (4) (Figure 10) and ANN-4/9/1, BPGD learning algorithm was chosen arbitrarily. Sigmoid function and linear function were used in sequence as the activation functions in the three-layer structure. The learning rate took the value of 0.003 while the momentum constant took the value of 0.9. For unit (5) (Figure 10), and ANN-4/3/2/1, RPROP training algorithm and TANH activation function were chosen. The parameters of the RPROP were $\eta^+ = 1.2$ and $\eta^- = -0.5$, as recommended in references [41,42]. Initial values of weight coefficients were chosen randomly from the range $[-0.5, 0.5]$. All of the inputs and outputs were normalized in the interval between 0 and 1 by linear scaling. The chosen number of learning epochs was 1000, to avoid overtraining. After 30 repetitions of the modeling process, the most successful result, which was rated by the lowest root mean squared error on single tests points, was chosen as relevant. The error of the adopted model is estimated as the mean RMSE of all the generated ANN models. Performance metric values of the adopted final data table numerical model of REE extraction process (Figure 13), are $\overline{RMSE}_{train} = 0.856$, for all of the training data sets and $\overline{RMSE}_{test} = 3.268$, for all the test single points.

The adopted final model (6), depended on four input variables, and its response surface could not be presented in a three-dimensional (3D) diagram. In order to visualize the prediction result, some 3D plots of the response surface were obtained by maintaining some selected factors constant. Some typical diagrams are shown in Figure 13a,b.

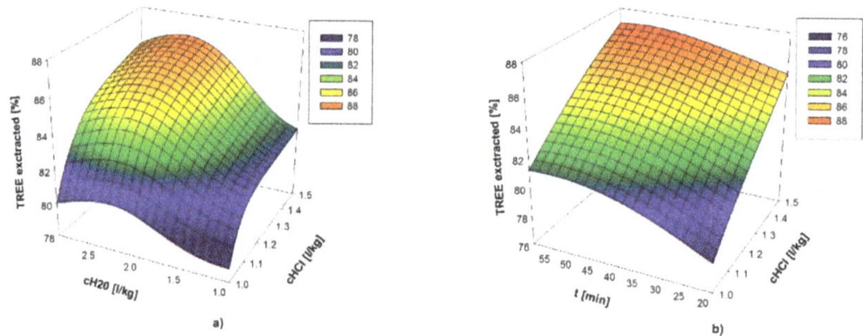

Figure 13. Final ANN data table numerical model for TREE extraction: (**a**) $T = 20\ ^\circ C = const.$, $t = 60$ min. $= const.$; (**b**) $T = 20\ ^\circ C = const.$, $cH_2O = 1.5$ L/kg $= const.$

6. Validation of the Model and Scale-up

6.1. Validation of REE Extraction Model and Obtained Optimal Process Regimes

In Section 4, the realized CCD plan for dry digestion and the leaching process, was presented. Using the obtained experimental results, mathematical models of the investigated process were created in MLR, SWR, and ANN forms. Final ANN data table numerical model for TREE extraction, given in Figure 13, was adopted as adequate. Based on this model the optimum intervals for the values of predictor variables are determined. These optimized intervals define the optimal regimes for the high performance of the REE extraction process. The optimal intervals for the parameters in stage I and stage II are: hydrochloric acid to concentrate ratio between 1.3 and 1.50, water to concentrate between 2.3–2.5 L/kg, leaching temperature between 20 °C and 25 °C (no external heating), and leaching time between 40 and 60 min. Under these conditions, the model predicted TREE extraction between 86% and 90%. However, considering the cost of REE extraction, in order to obtain and confirm the final values of optimal parameters that would be applied to the pilot scale-up system, some validation experiments were performed in the lab. The operation process was the same as the experiments in the adopted CCD plan. In accordance with the above, four additional experiments were carried out. Each of these four experiments (Table 5) was realized under the same leaching temperature—20 °C.

Table 5. The results of experiment checking.

Parameter	Units	B1	B2	B3	B4
HCl:Concentrate	L:kg	1.3:1	1.4:1	1.4:1	1.5:1
Water:Concentrate	L:kg	2.25:1	2.25:1	2.5:1	3:1
Leaching time	min	40	50	50	60
Predicted value		85.35%	87.90%	89.40%	89.40%
Actual value		87.65%	89.40%	89.70%	91.20%

The mean TREE extraction efficiency for these four experiments, under laboratory conditions, was 89.49% and mean RMSE was $RMSE_{ANN}$ = 1.65. The obtained results confirmed the adequacy of the adopted ANN model, which was the basis for selecting the optimal regime of the REE extraction technological procedure on the scale-up level.

6.2. Dry Digestion and Leaching Process Scale-Up

The laboratory results were validated at one newly built demonstration plant at the IME, RWTH Aachen in order to provide more detailed knowledge of the REE extraction process before industrialization application. Treatment of 500 kg of eudialyte concentrate was performed using two dry digestion reactors (as shown in Figure 14) under the optimal values of process parameters that were obtained using adopted ANN model (Figure 13), described in Section 6.1.

There were two reactors for dry digestion process at the demonstration plant and the capacity of each reactor was 40 L. When considering that the slurry was thick after mixing the acid and eudialyte concentrate, the handling capacity was controlled not to exceed a maximum of 6 kg concentrate per reactor. After performing dry digestion and leaching adding specified amounts of water and transferring the slurry to a large tank (200 L), leaching was then carried out at the temperature of around 20 °C. After neutralization with limestone until the pH reached 4.0, the slurry was pumped to a filter press and the separation of solid and liquid was achieved. The final REE enriched solution was obtained after washing the filter cake, and the composition of the REE enriched solution is shown in Table 6. The experiments were repeated 3–5 times per day. In total, 48 experiments were realized on the scale-up demonstration plant. The mean overall TREE extraction efficiency under adopted optimal regimes defined by model given in Figure 13 was 86.9%.

Figure 14. Treatment of eudialyte concentrate in two reactors at the Institute of Process Metallurgy and Metal Recycling (IME) demonstration plant, RWTH Aachen.

Table 6. Chemical composition of REE enriched solution II in pilot scale-up test.

Element	Concentration (g/L)	Element	Concentration (g/L)
Al	0.56	Ce	0.951
Ca	10.49	Pr	0.105
Fe	2.02	Nd	0.370
Mn	0.269	Sm	0.094
Nb	<0.0001	Gd	0.075
Zr	0.02	Dy	0.105
Hf	0.010	Y	0.601
Si	0.099	Yb	0.069
La	0.425	TREE	2.80

7. Discussion

7.1. REE Extraction Modeling and Optimal Regimes

The discussion related to the selection of significant predictors in the REE extraction process, the variation intervals of these predictors, the applied CCD experimental plan and the developed mathematical MLR, SWR, and ANN models, with their accuracy and properties, are given in Sections 4 and 5. The reliability of the ANN model is enhanced by utilizing the LOOCV method. ANN is a nonlinear model capable of incorporating the nonlinear effects of predictor variables and their combined effects on the response variables. This model is also capable to overcome the high levels of noise that characterize the REE extraction process. For these reasons, the adopted ANN model shows better performance and accuracy ($\overline{RMSE}_{test} = 3.268$, $\overline{RMSE}_{train} = 0.856$) in comparison with MLR ($RMSE_{(8)} = 4.668$, $R^2_{adj(8)} = 0.244$) and SWR ($RMSE_{(9)} = 4.487$, $R^2_{adj(9)} = 0.413$) models.

The validity of the adopted numerical ANN model that is shown in Figure 13 was confirmed under laboratory conditions (Table 5). Optimal variation intervals of the process parameters that predict the maximization of the TREE extraction efficiency from the eudialyte ore (Section 6.1) were determined. As previously described, the expected TREE extraction efficiency is in the range of 86 to 90%.

Tests at the IME RWTH Aachen demonstration plant, under optimal regimes, resulted in an overall average TREE extraction efficiency of 86.9%. The chosen parameter values offer the prevention of silica gel formation. Such results additionally confirm the feasibility of the proposed and designed REE extraction process from eudialyte concentrate and its potential for industrial application. The ratio between concentrate and hydrochloric acid is enough to dissolve rare earth elements from eudialyte concentrate. The leaching time with water between 40 and 60 min is shorter in comparison to the acid baking process and other hydrometallurgical processes under an atmospheric pressure.

Previous models and discussions have indicated that the leaching temperature does not have a significant effect on the TREE extraction efficiency. This fact is of great importance because it has a positive effect both on reducing the costs of REE extraction (no external heating) and on avoiding very complex problems that are related to the types of materials from which reactors and other equipment are made. Additionally, this may have a positive effect on avoiding problems related to corrosion at high temperatures.

To further explore the REE extraction process, additional, potentially significant predictors should be included in the analysis. The first possibly significant predictor, which will be considered in future research, is stirring rate. Potentially, this predictor can influence both the prevention of silica gel formation and the overall duration of the process.

7.2. Phase Changes during REE Extraction

During the REE extraction process there are also phase changes. The analysis and discussion of these changes, in the solid residue, after dry digestion of the eudialyte concentrate with HCl, leaching with water, filtration, and overnight drying, is given in the following text.

As can be seen from Figure 15, the particles of eudialyte leach residue had irregular sizes and the agglomeration of particles could be observed. Figure 16 reveals the mineral composition of the leach residue. Eudialyte residue represented Si-O rich residue phase produced in digestion process, it could not be dissolved into the solution so that the presence of Si in the leaching solution was very low. The SEM result (Figure 17) confirms the same morphological characteristics. It can be explained by the fact that siliceous precipitation formed after HCl treatment can inhibit the leaching of REE. In Figure 18, the REE elemental mapping of eudialyte concentrate and leach residue are given together for the comparison purpose.

The field scan image of eudialyte concentrate in Figure 18 reveals that REE were inhomogeneous and mainly existed in the eudialyte phase. After acid digestion and leaching, few REE was detected in the phases of eudialyte and altered eudialyte.

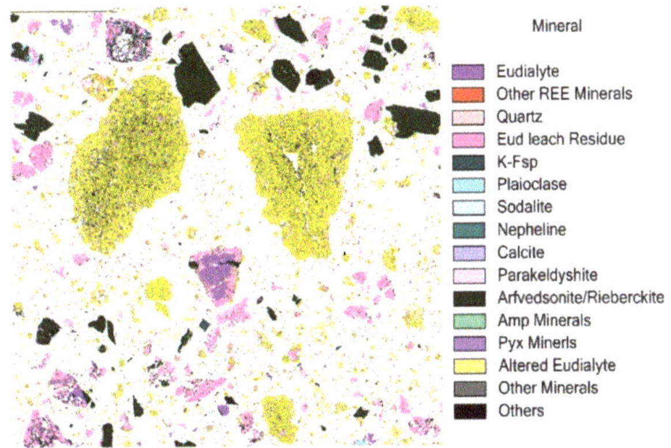

Figure 15. QEMSCAN analysis of eudialyte leach residue.

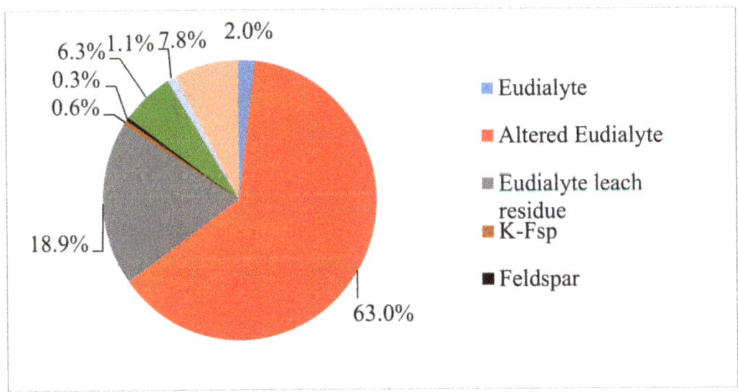

Figure 16. Mineral distribution of a eudialyte leach residue.

Figure 17. SEM image of a eudialyte leach residue.

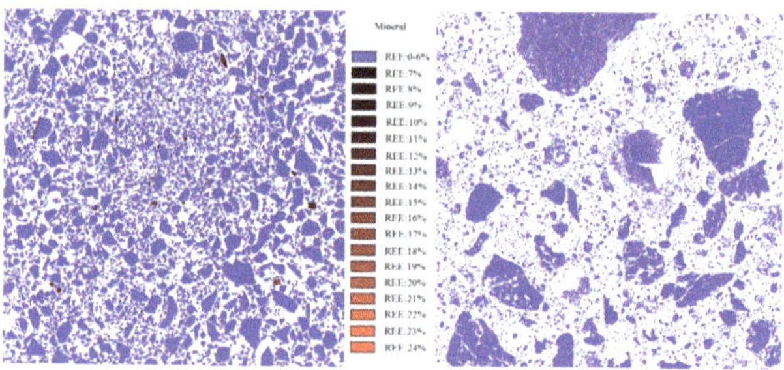

Figure 18. REE elemental mapping of eudialyte concentrate and leach residue by QEMSCAN.

8. Conclusions

Taking into account the state of the art regarding rare earth elements and the foreseeable, significant increase in the use of REE in various technological domains in the near future, the following assumptions have been made in this study: (a) the REE extraction process from eudialyte concentrate can be carried out at room temperature with a leaching time period that is not exceeding the leaching time periods used in other studies and with an acid concentration that is not higher than the acid concentrations used in other studies, (b) the REE extraction process can, on the basis of an optimal CCD plan, be modeled with a set of suitable MLR, SWR, and/or ANN models, (c) based on the most suitable mathematical model, optimal technological regimes can be determined, which allow for a satisfactory REE extraction efficiency, and (d) a scale-up demonstration plant can verify the feasibility and the efficiency of eudialyte concentrate treatment at low leaching temperatures.

The novelty in this study is the application of the DOE paradigm, the CCD optimal design of experiments, and the creation of a number of mathematical MLR, SWR, or ANN models, which more deeply describe the REE extraction process from eudialyte concentrate. Based on the realized CCD plan under laboratory conditions, various MLR, SWR, and ANN models of dry digestion and leaching process were created and used to predict the effects of various parameters such as hydrochloric acid to eudialyte concentrate ratio, water for leaching to eudialyte concentrate ratio, leaching temperature, and leaching time on the TREE extraction efficiency. The ANN model, which is based on LOOCV, was selected as the most suitable by the authors of this paper. ANN is a nonlinear model capable of incorporating the nonlinear effects of predictor variables and their combined effects on the response variables. This model is also capable to overcome the high levels of noise that characterize the REE extraction process. The reliability of the ANN model is improved with the use of the LOOCV method. For these reasons, the adopted ANN model shows better performance and in comparison with and SWR models. This approach for the extraction of rare earth elements during dry digestion and leaching process is missing in literature, and not present in previous work of Voßenkaul [24]. The research reported in this paper, also as a novelty, explored the feasibility and efficiency of the REE extraction process at room temperature on a scale-up demonstration platform that precedes future industrial application. Information on upscaling operations for treatment of eudialyte is also missing in overall literature.

In this study, it was confirmed that the proposed process is capable of achieving high REE recovery from an eudialyte concentrate during dry digestion with hydrochloric acid and leaching with water, avoiding silica gel formation and lowering operation costs (no heating for dry digestion and leaching). Based on the modeling results (as shown in Figure 13a,b), the optimal parameters in stage I and stage II are: hydrochloric acid to concentrate ratio between 1.3 and 1.50, water to concentrate between 2.3 and 2.5 L/kg, leaching temperature between 20 °C and 25 °C (no external heating), leaching time between 40 and 60 min. Finally, using the four parameters from the optimal range, the scale-up test was conducted at the IME-demonstration plant and the overall efficiency of the TREE extraction was 86.90%, which confirmed the feasibility and achieved good performance.

In comparison to the other process with sulphuric acid: Acid baking (250–300 °C; duration 2–4 h) [49] and Kvanefjeld process (80–95 °C, with a duration of approximately 16 h for the first and second leaching phase) [50], our REE extraction process is an environmentally friendly process, at room temperature, with a duration of approximately 2 h. During our research, we found that the leaching temperature has no significant influence on the increase of the extraction efficiency. Consequently, our proposed process is cheaper than other mentioned processes.

The obtained results are very promising, and will be used in our future work in order to propose and design a new dry digestion reactor (approximate volume 100 L) for an industrial application. This reactor might be used, not only for the treatment of a eudialyte concentrate, but also for a Bauxite residues (red mud) from the aluminum industry, what is mentioned in our newest publication in Nature [25]. Our future research will also take into consideration the stirring rate as one of the possibly important parameters regarding the avoidance of gel formation, leaching efficiency, and the duration of the REE extraction process.

Acknowledgments: The research leading to these results has received funding from the European Community's Seventh Framework Programme (Call identifier FP7-NMP-2012-LARGE-6) under grant agreement No. 309373. This publication reflects only the author's view, exempting the Community from any liability. Project website: www.eurare.eu. Authors are thankful to Geological Survey of Finland (GTK) for providing ore concentrate. One of the authors (Yiqian Ma) is grateful to the Chinese Government for providing a scholarship.

Author Contributions: Yiqian Ma and Srecko Stopic performed the experiments and wrote the paper; Srecko Stopic and Lars Gronen contributed reagents/materials/analysis tools; Yiqian Ma, Milovan Milivojevic and Bernd Friedrich analyzed the data. The optimization of process using Multiple Linear Regression (MLR), Stepwise Regression (SWR) and Artificial Neural Network was made by Milovan Milivojevic and Srdjan Obradovic.

Conflicts of Interest: The authors declare no conflict of interest.

References

1. Thomas, P.J.; Carpenter, D.; Boutin, C.; Allison, J.E. Rare earth elements (REEs): Effects on germination and growth of selected crop and native plant species. *Chemosphere* **2014**, *96*, 57–66. [CrossRef] [PubMed]
2. Morais, C.A.; Ciminelli, V.S.T. Process development for the recovery of high-grade lanthanum by solvent extraction. *Hydrometallurgy* **2004**, *73*, 237–244. [CrossRef]
3. Maestro, P.; Huguenin, D. Industrial applications of rare earths: Which way for the end of the century? *J. Alloys Compd.* **1995**, *225*, 520–528. [CrossRef]
4. Krishnamurthy, N.; Gupta, C.K. *The Rare Earths, Extractive Metallurgy of Rare Earths*; CRC Press: Boca Raton, FL, USA, 2015; pp. 1–84.

5. Hoshino, M.; Sanematsu, K.; Watanabe, Y. REE mineralogy and resources. In *Handbook on the Physics and Chemistry of Rare Earths*; Elsevier: New York, NY, USA, 2016; Volume 49, pp. 129–291.
6. McLemore, V.T. Rare earth elements (REE) deposits associated with great plain margin deposits (alkaline-related), southwestern united states and eastern mexico. *Resources* **2018**, *7*, 8. [CrossRef]
7. Möller, V.; Williams-Jones, A.E. A hyperspectral study (V-NIR-SWIR) of the Nechalacho REE-Nb-Zr deposit, Canada. *J. Geochem. Explor.* **2018**, *188*, 194–215. [CrossRef]
8. Alonso, E.; Sherman, A.M.; Wallington, T.J.; Everson, M.P.; Field, F.R.; Roth, R.; Kirchain, R.E. Evaluating rare earth element availability: A case with revolutionary demand from clean technologies. *Environ. Sci. Technol.* **2012**, *46*, 3406–3414. [CrossRef] [PubMed]
9. Golev, A.; Scott, M.; Erskine, P.D.; Ali, S.H.; Ballantyne, G.R. Rare earths supply chains: Current status, constraints and opportunities. *Resour. Policy* **2014**, *41*, 52–59. [CrossRef]
10. Goodenough, K.; Schilling, J.; Jonsson, E.; Kalvig, P.; Charles, N.; Tuduri, J.; Deady, E.; Sadeghi, M.; Schiellerup, H.; Müller, A. Europe's rare earth element resource potential: An overview of REE metallogenetic provinces and their geodynamic setting. *Ore Geol. Rev.* **2016**, *72*, 838–856. [CrossRef]
11. Balomenos, E.; Davris, P.; Deady, E.; Yang, J.; Panias, D.; Friedrich, B.; Binnemans, K.; Seisenbaeva, G.; Dittrich, C.; Kalvig, P. The EURARE project: Development of a sustainable exploitation scheme for Europe's Rare Earth Ore deposits. *Johns. Matthey Technol. Rev.* **2017**, *61*, 142–153. [CrossRef]
12. García, M.V.R.; Krzemień, A.; Del Campo, M.Á.M.; Álvarez, M.M.; Gent, M.R. Rare earth elements mining investment: It is not all about China. *Resour. Policy* **2017**, *53*, 66–76. [CrossRef]
13. Torró, L.; Proenza, J.; Aiglsperger, T.; Bover-Arnal, T.; Villanova-de-Benavent, C.; Rodríguez-García, D.; Ramírez, A.; Rodríguez, J.; Mosquea, L.; Salas, R. Geological, geochemical and mineralogical characteristics of REE-bearing Las Mercedes bauxite deposit, Dominican Republic. *Ore Geol. Rev.* **2017**, *89*, 114–131. [CrossRef]
14. Edahbi, M.; Benzaazoua, M.; Plante, B.; Doire, S.; Kormos, L. Mineralogical characterization using QEMSCAN® and leaching potential study of REE within silicate ores: A case study of the Matamec project, Québec, Canada. *J. Geochem. Explor.* **2018**, *185*, 64–73. [CrossRef]
15. Mikhailova, J.; Pakhomovsky, Y.A.; Ivanyuk, G.Y.; Bazai, A.; Yakovenchuk, V.; Elizarova, I.; Kalashnikov, A. REE mineralogy and geochemistry of the Western Keivy peralkaline granite massif, Kola Peninsula, Russia. *Ore Geol. Rev.* **2017**, *82*, 181–197. [CrossRef]
16. Deymar, S.; Yazdi, M.; Rezvanianzadeh, M.R.; Behzadi, M. Alkali metasomatism as a process for Ti–REE–Y–U–Th mineralization in the Saghand Anomaly 5, Central Iran: Insights from geochemical, mineralogical, and stable isotope data. *Ore Geol. Rev.* **2018**, *93*, 308–336. [CrossRef]
17. Davris, P.; Stopic, S.; Balomenos, E.; Panias, D.; Paspaliaris, I.; Friedrich, B. Leaching of rare earth elements from eudialyte concentrate by suppressing silica gel formation. *Miner. Eng.* **2017**, *108*, 115–122. [CrossRef]
18. Borst, A.M.; Friis, H.; Andersen, T.; Nielsen, T.; Waight, T.E.; Smit, M.A. Zirconosilicates in the kakortokites of the Ilímaussaq complex, South Greenland: Implications for fluid evolution and high-field-strength and rare-earth element mineralization in agpaitic systems. *Miner. Mag.* **2016**, *80*, 5–30. [CrossRef]
19. Anthony, J.W.; Bideaux, R.A.; Bladh, K.W.; Nichols, M.C. *Handbook of Mineralogy, Volume IV, Arsenates, Phosphates, Vanadates*; Mineralogical Society of America: Chantilly, VA, USA, 2000.
20. Johnsen, O.; Ferraris, G.; Gault, R.A.; Grice, J.D.; Kampf, A.R.; Pekov, I.V. The nomenclature of eudialyte-group minerals. *Can. Miner.* **2003**, *41*, 785–794. [CrossRef]
21. Zakharov, V.; Maiorov, D.; Alishkin, A.; Matveev, V. Causes of insufficient recovery of zirconium during acidic processing of lovozero eudialyte concentrate. *Rus. J. Non-Ferr. Met.* **2011**, *52*, 423–428. [CrossRef]
22. Lebedev, V. Sulfuric acid technology for processing of eudialyte concentrate. *Rus. J. Appl. Chem.* **2003**, *76*, 1559–1563. [CrossRef]
23. Lebedev, V.; Shchur, T.; Maiorov, D.; Popova, L.; Serkova, R. Specific features of acid decomposition of eudialyte and certain rare-metal concentrates from Kola peninsula. *Rus. J. Appl. Chem.* **2003**, *76*, 1191–1196. [CrossRef]
24. Voßenkaul, D.; Birich, A.; Müller, N.; Stoltz, N.; Friedrich, B. Hydrometallurgical processing of eudialyte bearing concentrates to recover rare earth elements via low-temperature dry digestion to prevent the silica gel formation. *J. Sustain. Metall.* **2017**, *3*, 79–89. [CrossRef]

25. Alkan, G.; Yagmurlu, B.; Cakmakoglu, S.; Hertel, T.; Kaya, Ş.; Gronen, L.; Stopic, S.; Friedrich, B. Novel Approach for Enhanced Scandium and Titanium Leaching Efficiency from Bauxite Residue with Suppressed Silica Gel Formation. *Nat. Sci. Rep.* **2018**, *8*, 5676. [CrossRef] [PubMed]
26. Vaccarezza, V.; Anderson, C. *Beneficiation and Leaching Study of Norra Kärr Eudialyte Mineral*; TMS Annual Meeting & Exhibition; Springer: Berlin, Germany, 2018; pp. 39–51.
27. Rivera, R.M.; Ulenaers, B.; Ounoughene, G.; Binnemans, K.; Van Gerven, T. Extraction of rare earths from bauxite residue (red mud) by dry digestion followed by water leaching. *Miner. Eng.* **2018**, *119*, 82–92. [CrossRef]
28. Milivojevic, M.; Stopic, S.; Friedrich, B.; Stojanovic, B.; Drndarevic, D. Computer modeling of high-pressure leaching process of nickel laterite by design of experiments and neural networks. *Int. J. Miner. Metall. Mater.* **2012**, *19*, 584–594. [CrossRef]
29. Milivojevic, M.; Stopic, S.; Stojanovic, B.; Drndarevic, D.; Bernd, F. Forward stepwise regression in determining dimensions of forming and sizing tools for self-lubricated bearings. *METALL* **2013**, *67*, 92–98.
30. Mata, J. Interpretation of concrete dam behaviour with artificial neural network and multiple linear regression models. *Eng. Struct.* **2011**, *33*, 903–910. [CrossRef]
31. Cohen, J.; Cohen, P.; West, S.G.; Aiken, L.S. *Applied Multiple Regression/Correlation Analysis for the Behavioral Sciences*; Routledge: Abingdon, UK, 2013.
32. Fisher, R.A. *The Design of Experiments*; Oliver and Boyd: Edinburgh, UK; London, UK, 1937.
33. Box, G.E.; Wilson, K.B. On the experimental attainment of optimum conditions. In *Breakthroughs in Statistics*; Springer: Berlin, Germany, 1992; pp. 270–310.
34. Dong, L.; Park, K.-H.; Zhan, W.; Guo, X.-Y. Response surface design for nickel recovery from laterite by sulfation-roasting-leaching process. *Trans. Nonferr. Met. Soc. China* **2010**, *20*, s92–s96.
35. Rumelhalt, D. *Learning Internal Representation by Error Propogation, Parallel Distributed Processing: Explorations in the Microstructure of Cognition (Vol. 1)*; MIT Press: Cambridge, MA, USA, 1986.
36. Drndarevic, D. Modeling and Optimization of Powder Metallurgy Process by Neural Networks. Ph.D. Thesis, University of Belgrade, Beograd, Serbia, 1996.
37. Law, R. Back-propagation learning in improving the accuracy of neural network-based tourism demand forecasting. *Tour. Manag.* **2000**, *21*, 331–340. [CrossRef]
38. Majdi, A.; Beiki, M. Evolving neural network using a genetic algorithm for predicting the deformation modulus of rock masses. *Int. J. Rock Mech. Min. Sci.* **2010**, *47*, 246–253. [CrossRef]
39. Yu, Z. *Feed-Forward Neural Networks and Their Applications in Forecasting*; University of Houston: Houston, TX, USA, 2000.
40. Heaton, J. *Programming Neural Networks with Encog 3 in c#*; Heaton Research, Inc.: St. Louis, MI, USA, 2015.
41. Riedmiller, M.; Braun, H. A Direct Adaptive Method for Faster Backpropagation Learning: The RPROP algorithm. In Proceedings of the IEEE International Conference on Neural Networks, San Francisco, CA, USA, 28 March–1 April 1993; pp. 586–591.
42. Riedmiller, M. Advanced supervised learning in multi-layer perceptrons—From backpropagation to adaptive learning algorithms. *Comput. Stand. Interfaces* **1994**, *16*, 265–278. [CrossRef]
43. Kingma, D.P.; Ba, J. Adam: A method for stochastic optimization. *arXiv*. 2014. arXiv.org e-Print archive. Available online: https://arxiv.org/abs/1412.6980 (accessed on 22 December 2014).
44. Milivojević, M. Methods of Development and Adaptation of Regression Models Based on Genetic Algorithms. Ph.D. Thesis, University of Kragujevac, Kragujevac, Serbia, 2016.
45. Pallant, J. *SPSS Survival Manual*, 3rd ed.; Mc Graw Hill: New York, NY, USA, 2007.
46. Team, R. *RR Development Core Team: A Language and Environment for Statistical Computing*; R Foundation for Statistical Computing: Vienna, Austria, 2012; ISBN 3-900051-07-0.
47. Hecht-Nielsen, R. Kolmogorov's mapping neural network existence theorem. In Proceedings of the International Conference on Neural Networks, San Diego, CA, USA, 21–24 June 1987; pp. 11–14.
48. McCaffrey, J. Understanding and using K-fold cross validation for neural networks. *Visual Studio Magazine*, 24 October 2013.

49. Sadri, F.; Nazari, A.M.; Ghahreman, A. A review on the cracking, baking and leaching processes of rare earth element concentrates. *J. Rare Earths* **2017**, *35*, 739–752. [CrossRef]
50. Krebs, D.; Furfaro, D. Processing of Rare Earth and Uranium Containing Ores and Concentrates. Patent WO 2014/082113 A1, 5 June 2014.

MDPI
St. Alban-Anlage 66
4052 Basel
Switzerland
Tel. +41 61 683 77 34
Fax +41 61 302 89 18
www.mdpi.com

Metals Editorial Office
E-mail: metals@mdpi.com
www.mdpi.com/journal/metals

www.ingramcontent.com/pod-product-compliance
Lightning Source LLC
LaVergne TN
LVHW070923170726
843515LV00005B/857

* 9 7 8 3 0 3 6 5 0 5 4 8 0 *